T4-AQK-713

DNA-BASED MOLECULAR CONSTRUCTION

Related Titles from AIP Conference Proceedings

633 Structural and Electronic Properties of Molecular Nanostructures: XVI International Winterschool on Electronic Properties of Novel Materials
Edited by Hans Kuzmany, Jörg Fink, Michael Mehring, and Siegmar Roth, November 2002, 0-7354-0088-1

591 Electronic Properties of Molecular Nanostructures: XV International Winterschool/Euroconference
Edited by Hans Kuzmany, Jörg Fink, Michael Mehring, and Siegmar Roth, November 2001, 0-7354-0033-4

590 Nanonetwork Materials: Fullerenes, Nanotubes, and Related Systems, ISNM 2001
Edited by Susumu Saito, Tsuneya Ando, Yoshihiro Iwasa, Koichi Kikuchi, Mototada Kobayashi, and Yahachi Saito, October 2001, 0-7354-0032-6

544 Electronic Properties of Novel Materials—Molecular Nanostructures: XIV International Winterschool, Euroconference
Edited by Hans Kuzmany, Jörg Fink, Michael Mehring, and Siegmar Roth, November 2000, 1-56396-973-4

492 Simulation and Theory of Electrostatic Interactions in Solution: Computational Chemistry, Biophysics, and Aqueous Solutions
Edited by Lawrence R. Pratt and Gerhard Hummer, November 1999, 1-56396-906-8

To learn more about these titles, or the AIP Conference Proceedings Series, please visit the webpage **http://proceedings.aip.org/proceedings**

DNA-BASED MOLECULAR CONSTRUCTION

International Workshop on
DNA-Based Molecular Construction

Jena, Germany 23–25 May 2002

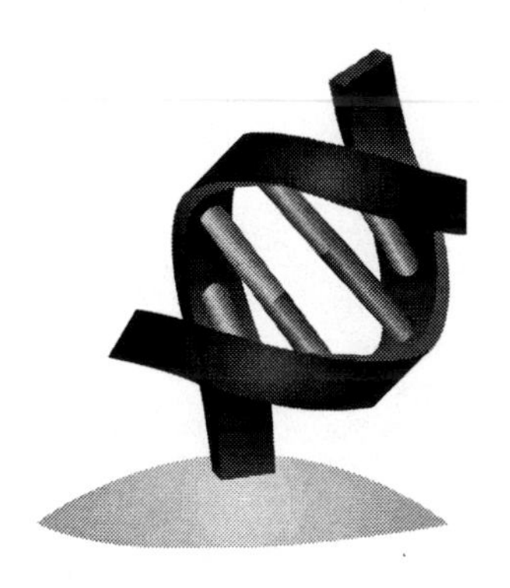

EDITOR
Wolfgang Fritzsche
IPHT Jena, Germany

SPONSORING ORGANIZATION
IPHT-Institut für Physikalische Hochtechnologie e.v. Jena

Melville, New York, 2002
AIP CONFERENCE PROCEEDINGS ■ VOLUME 640

Editor:

Wolfgang Fritzsche
IPHT-Institut für Physikalische Hochtechnologie
P.O. Box 100239
07702 Jena
GERMANY

E-mail: fritzsche@ipht-jena.de

Authorization to photocopy items for internal or personal use, beyond the free copying permitted under the 1978 U.S. Copyright Law (see statement below), is granted by the American Institute of Physics for users registered with the Copyright Clearance Center (CCC) Transactional Reporting Service, provided that the base fee of $19.00 per copy is paid directly to CCC, 222 Rosewood Drive, Danvers, MA 01923. For those organizations that have been granted a photocopy license by CCC, a separate system of payment has been arranged. The fee code for users of the Transactional Reporting Service is: 0-7354-0095-4/02/$19.00.

© 2002 American Institute of Physics

Individual readers of this volume and nonprofit libraries, acting for them, are permitted to make fair use of the material in it, such as copying an article for use in teaching or research. Permission is granted to quote from this volume in scientific work with the customary acknowledgment of the source. To reprint a figure, table, or other excerpt requires the consent of one of the original authors and notification to AIP. Republication or systematic or multiple reproduction of any material in this volume is permitted only under license from AIP. Address inquiries to Office of Rights and Permissions, Suite 1NO1, 2 Huntington Quadrangle, Melville, N.Y. 11747-4502; phone: 516-576-2268; fax: 516-576-2450; e-mail: rights@aip.org.

L.C. Catalog Card No. 2002113101
ISBN 0-7354-0095-4
ISSN 0094-243X
Printed in the United States of America

CONTENTS

APPENDICES

Preface

Molecular nanotechnology is an emerging technology that promises a new quality in materials and devices based on ultimately small units. Visionary thoughts of miniaturization, e.g. by Feynman and Drexler, opened a new world of scientific ideas. On the experimental side, the 1980s witnessed the development of methods for single-molecule characterization and manipulation, mainly based on optical (optical tweezers) and scanning probe techniques. In the 1990s, these techniques were further enhanced and complemented by other developments (e.g., soft lithography). The combination of these methods led at the turn of the century to a new quality in bringing the single-molecule experiments from solution onto surfaces, resulting, for example, in electronic circuits using single molecules.

From the beginning, DNA enjoyed special attention as material for molecular construction. The pioneering work of Seeman demonstrated a wide range of DNA constructions, later complemented by other groups in nanotechnology and by advances in DNA techniques driven by the progress in genomics.

At this stage of the science, the workshop "DNA-based molecular construction" brings together the interesting development in the different fields of molecular nanotechnology featuring DNA either as subject of studies or as a molecular building block. To the best of our knowledge, this was the first meeting to solely focus on DNA-nanotechnology. The interdisciplinary background of the participants in physics, chemistry, and biology provided a comprehensive account of the state-of-the-art and ensured the broadest exploration. It also highlighted the trends for further research in this promising field.

Wolfgang Fritzsche

Sponsors

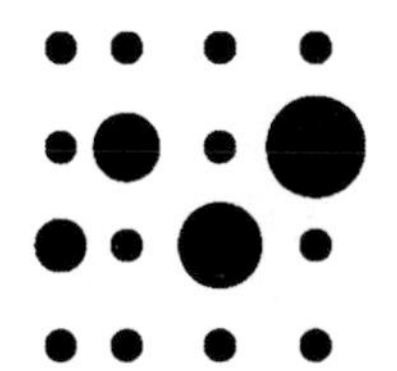

http://www.volkswagenstiftung.de

http://www.ipht-jena.de

http://www.bmbf.de/

http://www.dewb-vc.com/

http://www.vci.de/fonds/

http://www.stift-thueringen.de/

Scientific committee

Cees Dekker (Delft)

Fritz Eckstein (Göttingen)

Wolfgang Fritzsche (Jena)

Christine Keating (Penn State)

Christof Niemeyer (Bremen)

Nadrian Seeman (New York)

DNA AT SURFACES

Integrating DNA With Semiconductor Materials: Bio-inorganic Hybrid Devices

Andrew R. Pike[*], Lars H. Lie[*], Samson N. Patole[*], Lyndsey C. Ryder[*], Bernard. A. Connolly[†], Benjamin R. Horrocks[*] and Andrew Houlton[*].

[*]*School of Natural Sciences, University of Newcastle upon Tyne, Bedson Building, Newcastle upon Tyne, NE1 4RU.*

[†]*School of Cell and Molecular Biosciences, University of Newcastle upon Tyne, The Medical School, Newcastle upon Tyne, NE2 4HH.*

Abstract. A method for the automated solid-phase synthesis of DNA on a semiconductor chip with the potential for photolithography to fabricate hybrid electronic-DNA devices was developed. The on-chip oligonucleotide synthetic quality was comparable to standard CPG supports as confirmed by HPLC and gel electrophoresis. Enzymatic manipulation of the immobilised ssDNA was possible by radiolabelling with T4 polynucleotide kinase. Spatial control, afforded by photolithography, was visualised by phosphorimaging radiolabelled dsDNA. The charge transfer properties of DNA were investigated by the association of $Ru[(NH_3)_6]^{3+}$ with the phosphate backbone and by intercalation with redox active methylene blue. Additionally ferrocene modified nucleosides were incorporated into oligonucleotides to act as electronic mediators for charge transfer. Initial investigations into the effect of the redox group on the nucleobase indicated their potential for use as bioelectronic building blocks for incorporation into silicon based molecular systems.

INTRODUCTION

Molecular electronics has become an area of considerable interest ever since the physical and economic limitations of present day microelectronic technologies have become more and more apparent [1]. In 1965 Gordon Moore suggested that the density of components on an IC would double every year [2]. By simple extrapolation, device features will be of the nanometer scale by the year 2010. Up until now Moore's Law has more or less held, however, recently there is a slowing down of this miniaturization process as devices have reached the sub-micron scale. At the same time advances in the bio-sciences have been rapid especially in the sequencing of the human genome. Therefore, as the "world" of the electronics engineer appears to be shrinking, bio-molecules are taking centre stage in multidisciplinary research projects the world over. As a result, an overlap between these two areas has led to the development of numerous new technologies, including nanotechnology. Unsurprisingly DNA has played a key role in recent advances due to its programmable nanometer-scale dimensions and its highly characteristic supramolecular behavior, making it ideal for assembling molecular-based components [3-5]. Furthermore its ease of synthesis, either chemically, using solid-phase methods, or enzymatically

CP640, *DNA-Based Molecular Construction: International Workshop,* edited by W. Fritzsche
© 2002 American Institute of Physics 0-7354-0095-4/02/$19.00

allows for integration with chip technology. DNA has been used to form a range of organized nanostructures and even micron-scale interconnects [3-5]. Often DNA is attached to a surface, either a bulk solid, as in the case of metallic electrodes, or to nanoscopic particles, such as gold colloids [6-10]. Surprisingly, however, there have been very few reports of DNA-semiconductor interfaces [11-13]. Semiconductor silicon will continue to be the material of choice in electronic devices for the foreseeable future, and we therefore have investigated its integration with DNA [13]. We envision that simple and direct interconnections between silicon chips and bio-molecules such as DNA will be fundamental in the design of future molecular electronic devices [14]. Here the direct covalent attachment of DNA to the semiconductor surface, the enhancement of the potential conductivity of the DNA duplex and the patterning of such silicon-biomolecule junctions are described.

Modification Of Semiconductor Silicon

During the fabrication of a silicon chip the surface of a silicon wafer is patterned by photolithographic techniques. Pattern resolution is determined by the wavelength of light used to expose the native silicon surface through a layer of photoresist. It is this requirement for light of shorter and shorter wavelengths and the subsequent development of electron beam lithography that is making conventional chip-fabrication economically prohibitive at the nano-scale. The exposed regions are then etched to yield a hydride surface which is typically doped before the process is repeated and metal links inserted to build up the complex circuitry of the silicon chip. The step of interest to us is the etching process, as it leaves a hydrogen terminated silicon surface [15]. STM imaging shows that the surface has a step and terrace structure [16]. In conditions where the NH_4F has been de-oxygenated by purging with nitrogen, the terraces are atomically flat. The terrace width is in the range 40-50 nm depending on the miscut angle of the particular wafer, and the step height is less than 1 nm. The nature of the hydride surface depends on the orientation of the silicon crystal, <100> and <111> wafers upon etching yield di- and mono-hydride surfaces respectively, see Figure 1.

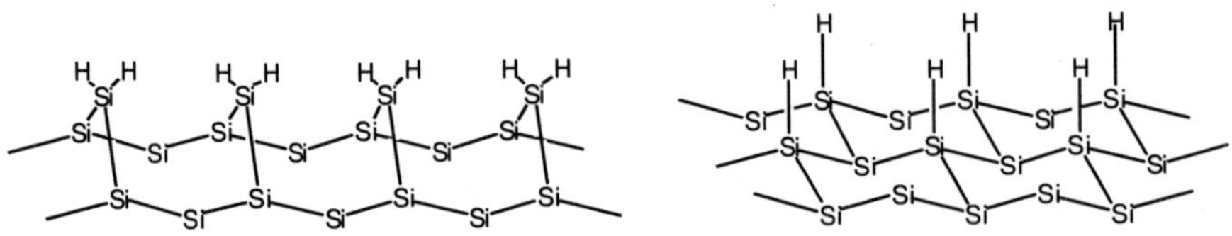

Si(100)

Si(111)

FIGURE 1. Schematic representation of the di-and mono-hydride surfaces after the etching of native silicon oxide wafers, <100> and <111>, by HF or 40% $NH_4F(aq)$.

FIGURE 2. Comparison of single crystal (left), and porous silicon (right) surfaces.

In addition to the flat single crystal hydrogen terminated surfaces, electrochemical etching at a fixed current density in HF gives a hydride surface with porous structure [17]. The size and nature of the pores can be controlled by changing the concentration of HF and the current density used in the etching process. After etching in a locally constructed electrochemical cell a darker circular region of porous silicon hydride can be clearly seen. The color of the porous region may be black, brown, green, or yellow according to the etch conditions employed. Figure 2 shows a comparison of the shiny and smooth single crystal surface with the nearly black surface of porous silicon. The porous micro-structure is not simple physically or chemically, and is terminated in a mixture of mono-, di- and tri-hydride species. The increased surface area of porous silicon over single crystal silicon makes it much easier to investigate the nature of the surface by FTIR. Therefore modifications performed at a single crystal surface can be monitored by looking at the IR spectrum of its porous counterpart, making it a convenient method to follow silicon wafer modification through chemistry of the Si-H surface.

Organic Reactions At Hydrogen Terminated Silicon Surfaces

We consider molecular chemistry at bulk semiconductor materials a crucial step in the development of molecular-based electronics. Chemistry that allows the molecular functionalisation of silicon has been an area of growing interest [18, 19]. It has been shown that the hydrogen-terminated layer, formed during silicon wafer processing, is reactive to molecular species resulting in covalently bonded monolayers. The hydride surface is extremely useful for the preparation of monolayers because the Si-H can mimic the reactivity of molecular hydrosilanes. The functional group chemistry of this surface has been explored and Si-O-C, Si-N or Si-C bonded monolayers have all been realised [20-26]. A variety of reactions can be performed and a brief summary is shown in Figure 3. We have shown that hydrosilation reactions work well for 1-alkenes and 1-alkynes as well as alcohols, giving well-ordered monolayers, even when prepared from dilute solutions in aromatic organic solvents [22-26]. Si-C bonded layers are extremely robust and in addition to reactions with alkenyl and alkynyl groups can be prepared by reactions with a variety of other reagents including organolithium and Grignard reagents [27, 28]. We have primarily focused on the reaction of alkenes to form stable Si-C bonded monolayers, which compared to the highly air- and moisture-sensitive nature of alkyl-lithium and Grignard reagents are relatively straightforward to handle. For example, by refluxing a freshly etched silicon wafer in 0.02 M alkene in toluene for 16 hours an alkyl monloayer can be easily formed.

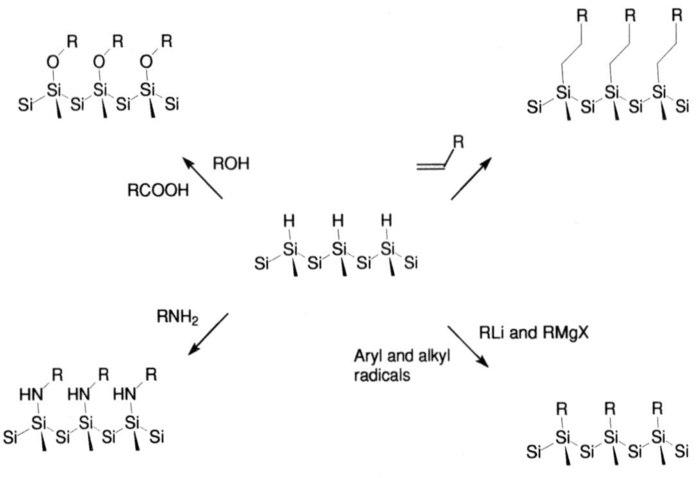

FIGURE 3. Summary of the organic reactions possible at silicon-hydrogen surfaces.

In order to form a monolayer with an exposed functional group a bifunctional molecule is required, one group reacting with the silicon hydride surface leaving the second free for further manipulation. The criteria of a molecular interconnect for the covalent attachment of DNA to semiconductor silicon is that the compound should have both a terminal hydroxyl site for DNA synthesis, and also a group reactive towards Si-H. The simplest class of such compounds is the alken-1-ols. However, the reactivity of both unsaturated C=C linkages and alcohols to the hydride surface is problematic, and protection of the hydroxyl group is necessary, Figure 4. The standard protection method of the 5'-OH of nucleosides is by the dimethoxytrityl (DMT) group. Therefore reaction of the primary alcohol of ω-undecen-1-ol with dimethoxytrityl chloride yielded the desired bifunctional connector, see Figure 4. Hydrosilation at the alkenyl group by the hydrogen-terminated surface gives a DMT-protected alcohol monolayer which replaces the commercial silica supports for DNA synthesis [13].

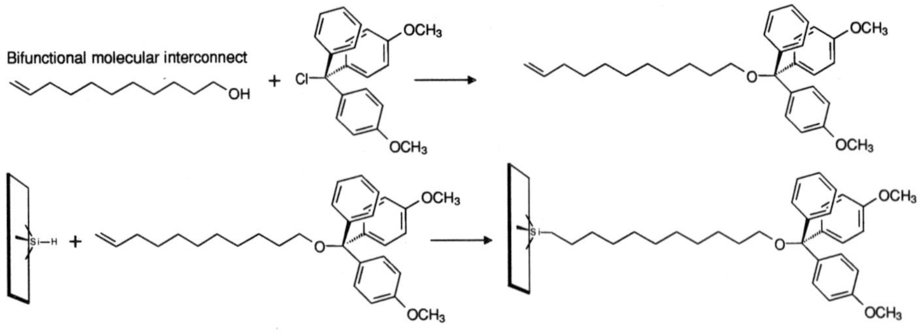

FIGURE 4. Protection of the terminal hydroxyl of an alken-1-ol to give a DMT protected monolayer on semiconductor silicon suitable for use as a solid support for automated DNA synthesis.

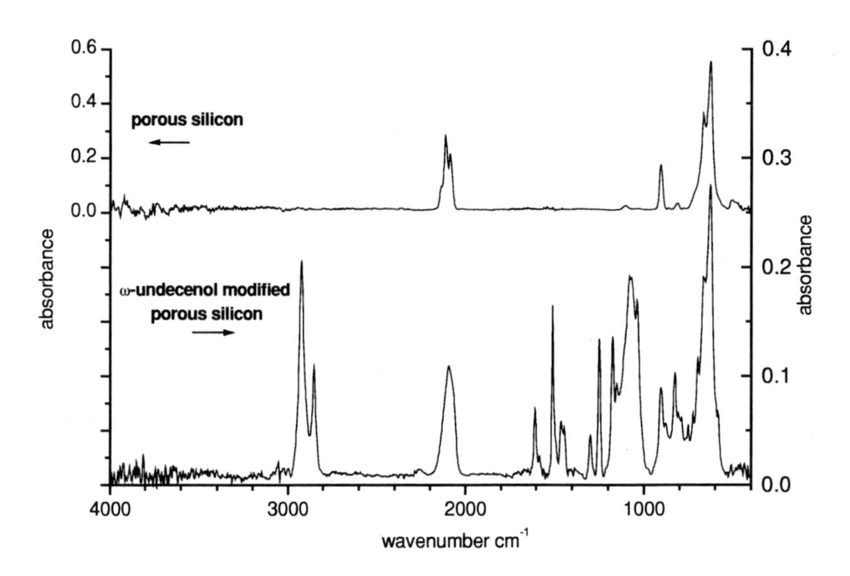

FGURE 5. FTIR spectra of porous silicon before and after alkylation with DMT-protected ω-undecenol.

FTIR spectroscopy of a porous silicon wafer before and after reaction with DMT-protected ω-undecen-1-ol was performed and the spectra are shown in Figure 5. Bands corresponding to $\nu(CH_2)$ at 2855 and 2927 cm^{-1} and the concurrent lack of the alkenyl C=C stretch band confirm alkylation. However the presence of a Si-H stretching band at 2109cm^{-1} indicated that the coverage was not complete, which is expected due to the nature of the porous surface. The modified silicon wafer is now ready to undergo solid-phase DNA synthesis at the protected alcohol monolayer in an automated synthesizer.

DNA Synthesis On Semiconductor Silicon

Automated solid-phase DNA synthesis uses phosphoramidite chemistry and is a quick and convenient method for building oligonucleotides of any desired base sequence and length. Through a cycle of reactions an oligonucleotide is built up base by base at a fixed support. By replacing the standard silica beads of a commercial column with our DMT-protected silicon wafer (either single crystal or porous) we could build directly onto the semiconductor surface a sequence of single stranded DNA (ssDNA) [13]. Modification of the standard protocol to avoid the harsh conditions of the final de-protection step was necessary, and so Ultramild bases were used. This adaptation meant that a 20 minute treatment in anhydrous methylamine was sufficient to remove the base-protecting groups [29]. To establish that the ssDNA was of good quality a cleavable sulfonyl ether linkage between the alkyl monolayer and the oligonucleotide was inserted. Analysis by HPLC, Figure 6, and by gel electrophoresis

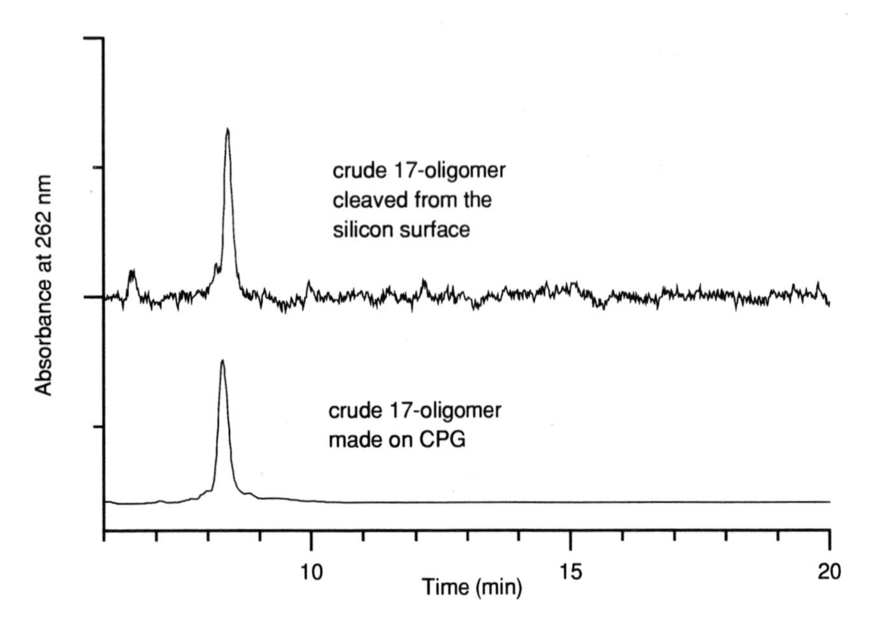

FIGURE 6. Comparison of HPLC data of crude oligomer synthesized on a CPG column and on a modified single crystal silicon wafer.

showed that our modified silicon wafer gave DNA oligomers of comparable quality to commercial silica (CPG) supports [13].

Having shown that the on-chip synthesis was effective we were interested to see if the silicon-DNA surfaces would undergo hybridization. Therefore we prepared patterned ssDNA-wafers using photolithographic methods [30]. UV light was used to develop a pattern on the silicon oxide surface through a film of positive photoresist. The exposed regions were then etched as normal, alkylated and DNA synthesis performed as already described. These ssDNA surfaces were treated with a solution containing the radiolabelled complementary oligonucleotide. Phosphorimages of the double stranded DNA (dsDNA) surfaces revealed the same pattern as the photolithographic mask, confirming hybridization. Successful radiolabelling of the terminal 5'-OH with [γ-^{32}P]ATP by T4 polynucleotide kinase established that the surface-bound oligonucleotide strands could be manipulated enzymatically [13]. The ability for enzymes to process the surface-bound DNA opens up numerous possibilities for further biochemical modifications of the surface structure.

The electrochemical quantitation of DNA surface coverage is possible due to the conductive properties of the semiconductor substrate. Potential step chronoamperometric measurements on $[Ru(NH_3)_6]^{3+}$ associated to the phosphate backbone allowed quantification of the ssDNA surface coverage. Intercalation of redox active methylene blue into duplex structures further demonstrated charge transfer through the surface-bound dsDNA. Methylene blue discriminates between ssDNA and dsDNA and only binds to the latter. Surface coverage was found to be 3.2 x 10^{12} and 1.1 x10^{12} molecules cm^{-1} for ssDNA and dsDNA respectively [13]. This suggests that 30% of the ssDNA undergoes hybridisation under the conditions employed. The data shows that redox active groups can be non-covalently incorporated into both ssDNA and dsDNA at silicon interfaces. In parallel to work on DNA attached to silicon we have also investigated enhancing the electron transfer properties of DNA itself, by chemically modifying DNA.

Hybrid Biomolecules: Modified DNA

The modification of DNA with metal containing groups has been widespread [31-40]. Much of the research has focused on the biological activity of such derivatives [35-37] and for investigating DNA-mediated charge transport processes [38-40].We have been interested in the functionalization of nucleobases with redox groups [41]. The pyrimidines have been commonly modified at C-5 with a variety of functionalities, in particular ferrocene [42-45]. Ferrocene is a one electron redox group and by connecting it through a covalent linkage to the thymine ring we envisage it to behave as an electronic mediator for charge transfer.

We synthesized a series of ethyl-, vinyl- and ethynylferrocenyl C5-modified-thymidine derivatives [44]. The extent of charge delocalization into the nucleobase in the oxidized ferrocenium counterpart, and the degree this was influenced by the nature of the covalent linkage were assessed using density functional theory (DFT) calculations and X-ray crystallography [44]. The results revealed that there is considerable interaction between the nucleobase and the redox-active ferrocenyl group. Delocalization occurred across all three linkages irrespective of the degree of unsaturation. The synthesis and purification of the ethynyl ferrocenyl thymidine nucleoside was the most straightforward due to isomerisation problems with the vinyl analogue. Ideally we wanted to be able to incorporate the redox active nucleoside in any number and at any position into an oligonucleotide by automated solid-phase synthesis, see Figure 7. Therefore the fully protected ethynyl ferrocenyl thymidine phosphoramidite was synthesized by standard protocols. After each synthetic step the intermediates were all electro-active as demonstrated by cyclic voltammetry and differential pulse voltammetry. Furthermore the 20-mer oligonucleotide, 5'<TATCGTATCGXATCGTATCG>3, where X is ethynyl ferrocenyl thymidine, had a redox peak at 282mV vs. Ag quasi reference electrode (QRE). This was considerably lower than the ferrocenyl monomer at 395mV which is explained by the proximity of the negatively charged phosphate backbone to the redox centre.

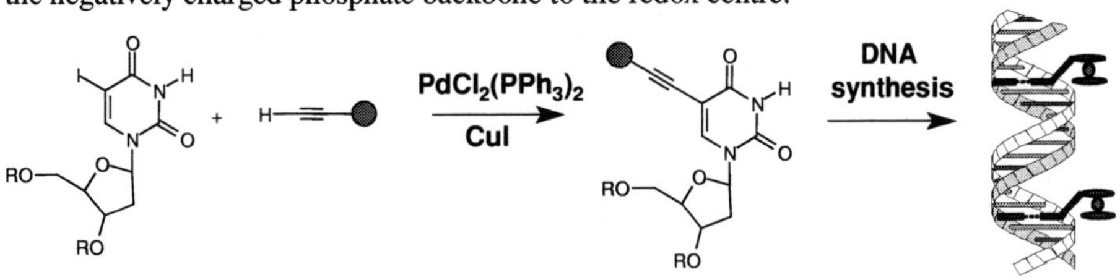

FIGURE 7. Schematic representation of the reaction of ethynyl ferrocene with 5-iodo thymidine under Sonogashira conditions to give the modified redox active nucleoside which is subsequently incorporated into DNA by automated solid phase synthesis.

Analysis by HPLC of the 20-mer oligonucleotide after digestion by a cocktail of snake venom phophodiesterase and alkaline phosphatase, indicated that 50% of the ferrocenyl nucleoside had undergone isomerisation during DNA synthesis [44]. This was in contrast to previous attempts to synthesize electrochemically active ferrocenyl-containing oligomers [45]. Therefore the methodologies developed here show promise for use in silicon-(bio)molecular hybrid electronics.

Bio-inorganic Hybrid Devices

The possibilities for the covalent assembly of hybrid devices as demonstrated here may prove useful in the construction of molecular-based electronics. However we have also started to investigate the concept of non-covalent bio-inorganic hybrid devices. Surprisingly, there have been few reports on the integration of molecular-based conductors such as polypyrrole, with silicon substrates [46, 47]. The use of non-covalent bonding based upon DNA-polypyrrole interactions can also be used to fabricate silicon-biomolecule-conducting polymer interfaces. As mentioned previously, the available dimensions and linear structure of DNA make it well suited as a molecular-based interconnect. Therefore preliminary experiments into the electro-deposition of polypyrrole onto a silicon/DNA junction were performed. DNA was synthesized onto exposed regions of a photolithographic pattern on a silicon wafer in the manner previously described. The modified sample was then used as an electrode to deposit films of polypyrrole. This was done by electrochemical oxidation of 5 % v/v pyrrole solution in 0.1 M $TBAPF_6$/MeCN by cycling the potential between -0.5 and 1.5 V vs Ag QRE under illumination by a standard tungsten 100 W lamp. Over a couple of cycles the inverse pattern of the lithographic mask was "developed" as the pyrrole polymerized and formed a black film at the silicon-biomolecule interface, see Figure 8.

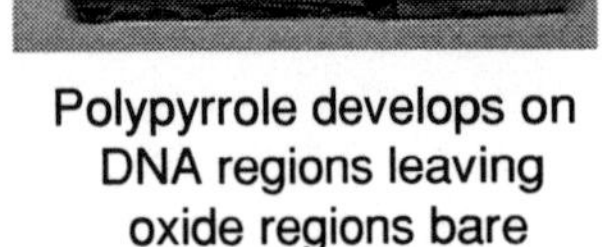

FIGURE 8. Si/DNA/polypyrrole junctions are formed by the electro-oxidation of pyrrole in 0.1 M $TBAPF_6$/MeCN by cyclic voltammetry using a DNA modified silicon wafer as working electrode, a silver wire reference electrode and tungsten wire counter electrode.

CONCLUSION

The possibilities for non-covalent assembly of hybrid devices at silicon semiconductor surfaces as demonstrated here may prove useful in the construction of molecular-based electronics. We have successfully integrated DNA with single crystal and porous silicon. Bio-manipulation through hybridization and enzymatic action as well as preliminary electrochemical investigations show that DNA-silicon surfaces hold potential as a building block for bio-inorganic hybrid devices. There is also the additional possibility for modifying DNA with electron mediators such as ferrocene and for templating polypyrrole at individual DNA-scaffolds. These concepts are currently being further explored in our laboratories.

ACKNOWLEDGMENTS

This work was in part funded by the EPSRC, the BBRSC, and the Research Council of Norway.

REFERENCES

1. Jortner, J., and Ratner, M., *Molecular Electronics*, Blackwell Science Ltd., Malden MA, 1997.
2. Moore, G. E., *Electronics*, **38**, 114-117, (1965)
3. Alivisatos, A. P., Johnsson, K. P., Peng, X. G., Wilson, T. E., Loweth, C. J., Bruchez, M. P., and Schultz, P. G., *Nature*, **382**, 609-611, (1996).
4. Seeman, N. C., *Angew. Chem., Int. Ed.*, **37**, 3220-3238, (1998).
5. Mirkin, C. A., *Inorg. Chem.*, **39**, 2258-2272, (2000).
6. Kasumov, A. Y., Kociak, M., Gueron, S., Reulet, B., Volkov, V. T., Klinov, D. V., and Bouchiat, H., *Science*, **291**, 280-282, (2001).
7. Niemeyer, C. M., *Angew. Chem., Int. Ed.*, **40**, 4128-4158, (2001).
8. Porath, D., Bezryadin, A., de Vries, S., and Dekker, C., *Nature*, **403**, 635-638, (2000).
9. Braun, E., Eichen, Y., Sivan, U., and Ben-Yoseph, G., *Nature*, **391**, 775-778, (1998).
10. Taton, T. A., Mirkin, C. A., and Letsinger, R. L., *Science*, **289**, 1757-1760, (2000).
11. Strother, T., Hamers, R. J., and Smith, L. M., *Nucleic Acids Res.*, **28**, 3535-3541, (2000).
12. Strother, T., Cai, W., Zhao, X. S., Hamers, R. J., and Smith, L. M., *J. Am. Chem. Soc.*, **122**, 1205-1209, (2000).
13. Pike, A. R., Lie, L. H., Eagling, R. A., Ryder, L. C., Patole, S. N., Connolly, B. A., Horrocks, B. R., and Houlton, A., *Angew. Chem. Int. Ed.*, **41**, 615-617, (2002).
14. Pike, A., Horrocks, B., Connolly, B., and Houtlon, A., *Aust. J. Chem.*, **55**, 191-194, (2002).
15. Wolf, S.,and Tauber, R. N., *Silicon Processing for the VLSI Era*, Lattice Press, Sunset Beach, 1986.
16. Wade, C. P., and Chidsey, C. E. D., *Appl. Phys. Lett.*, **71**, 1679-1681, (1997).
17. Canham, L.T., Houlton, M. R., Leong, W. Y., Pickering, C., Keen, J. M., *J. Appl. Phys.*, **70**, 422-431, (1991).
18, Linford, M. R., and Chidsey, C. E. D., *J. Am. Chem. Soc.*, **115**, 12631-12632, (1993).
19. Linford, M. R., Fenter, P., Eisenberger, and P. M., Chidsey, C. E. D., *J. Am. Chem. Soc.*, **117**, 3145-3155, (1995).
20. Bergerson, W. F., Mulder, J. A., Hsung, R. P., and Zhu, X. Y., *J. Am. Chem. Soc.*, **121**, 454-455, (1999).
21. Buriak, J. M., *Chem. Commun.*, 1051-1060, (1999).
22. Bateman, J. E., Horocks, B. R., and Houlton, A., *J. Chem. Soc. Faraday Trans.*, **93**, 2427-2431, (1997).
23. Cleland, G., Horrocks, B. R., and Houlton, A., *J. Chem. Soc. Faraday Trans.*, **91**, 4001-4003, (1995).

24. Bateman, J. E., Eagling, R.D. ,Worrall, D. R., Horrocks, B. R., and Houlton, A., *Angew. Chem. Int. Ed.*, **37**, 2683-2685, (1998).
25. Eagling, R. D., Bateman, J. E., Goodwin, N. J., Henderson, W., Horrocks, B. R., and Houlton, A., *J. Chem. Soc. Dalton Trans.*, 1273-1276, (1998)
26. Bateman, J. E., Eagling, R. D., Horrocks, B. R., and Houlton, A., *J. Phys. Chem. B*, **104**, 5557-5565, (2000).
27. Song, J. H., and Sailor, M. J., *J. Am. Chem. Soc.*, **120**, 2376-2381, (1998).
28. Kim, N. Y., and Laibinis, P. E., *J .Am. Chem. Soc.*, **120**, 4516-4517, (1998).
29. Reddy, M. P., Hanna, N.B., and Farooqui, F., *Nucleosides Nucleotides*, **16**, 1589-1598, (1997).
30. Bard, A. J., *Integrated Chemical Systems A Chemical Approach to Nanotechnology*, John Wiley & Sons, Inc., New York, NY, 1994.
31. Magda, D., Crofts, S., Lin, A., Miles, D., Wright, M., and Sessler, J. L., *J. Am. Chem. Soc.*, **119**, 2293-2294, (1997).
32. Rack, J. J., Krider, E. S., and Meade, T. J., *J. Am. Chem. Soc.*, **122**, 6287-6288, (2000).
33. Lewis, F. D., Helvoight, S. A., and Letsinger, R. L., *Chem Commun.*, 327-328, (1999).
34. Weizman, H., and Tor, Y., *J. Am. Chem. Soc.*, **123**, 3375-3376, (2001).
35. Meunier, P., Ouattara, I., Gautheron, B., Tirouflet, J., Camboli, D., and Besancon, J., *Eur. J. Med. Chem.*, **34**, 351-362, (1991).
36. Dreyer,G. B., and Dervan, P. B., *Biochemistry*, **24**, 968-972, (1985).
37. Sigman, D. S., *Acc. Chem. Res.*, **19**, 180-186, (1986).
38. Holmlin, R. E., Dandliker, P. J., and Barton, J. K., *Angew. Chem. Int. Ed.*, **36**, 2715-2730, (1997)
39. Meade, T. J., and Kayyem, J. F., *Angew. Chem. Int. Ed*, **34**, 352-354, (1995).
40. Meggers, E., Kusch, D., and Giese, B., *Helv. Chim. Acta.*, **80**, 640-652, (1997).
41. Price, C., Aslanoglu, M., Isaac, C. J., Elsegood, M. R. J., Clegg, W., Horrocks, B. R., and Houlton, A., *J. Chem. Soc. Dalton Trans.*, 4115-4120, (1996).
42. Yu, C. J., Wan, Y., Yowanto, H., Li, J., Tao, C., James, M.D., Tan, C. L., Blackburn, G. F., and Meade, T. J., *J. Am. Chem. Soc.*, **123**, 11155-11161, (2001).
43. Khan, S. I., and Grinstaff, M. W., *J. Am. Chem. Soc.*, **121**, 4704-4705, (1999).
44. Pike, A. R., Ryder, L.C., Horrocks, B.R., Clegg, W., Elsegood, M. R. J., Connolly, B. A., and Houlton, A., *Chem. Eur. J.* **8**, 2891-2899, (2002).
45. Yu, C. J., Yowanto, H., Wan, Y. J., Meade, T. J., Chong, Y., Strong, M., Donilon, L. H., Kayyem, J. F., Gozin, M., and Blackburn, G. F., *J. Am. Chem. Soc.*, **122**, 6767-6768, (2000).
46. Kim, N. Y., and Laibinis, P.E., *J. Am. Chem. Soc.*, **121**, 7162-7163, (1999).
47. Jeon, N. L., Choi, I. S., Whitesides, G. M., Kim, N. Y., Laibinis, P. E., Harada, Y., Finnie, K. R., Girolami, G. S., and Nuzzo, R. G., *Appl. Phys. Lett.*, **75**, 4201-4203, (1999).

Optical Detection Of DNA Constructs Based On Nanoparticles And Silver Enhancement

Guo-Jun Zhang, Robert Möller, Andrea Csáki, Wolfgang Fritzsche

Institute for Physical High Technology Jena, Germany, Biotechnical Microsystems Department
fritzsche@ipht-jena.de

Abstract. Nanoparticle-labeling was recently introduced for probing immobilized DNA by scanning force microscopy or optical detection. The optical detection has the potential of high parallelization in combination with miniaturization, thereby enabling a high sample throughput. However, a quantification of the optical signal and a correlation of this signal with the surface density of bound nanoparticles are needed. We will demonstrate the application of the silver enhancement procedure for a signal amplification to extend the dynamic range of the method. The specificity of the enhanced nanoparticle-labeling will be shown, and the influence of the surface density of immobilized molecules on the signal is studied. The results confirm that the proposed detection scheme is suitable for an application in molecular nanotechnology for the characterization of DNA-modified surfaces.

INTRODUCTION

DNA-based molecular construction is based on hybridization as a highly specific DNA-DNA-interaction. This effect allows the connection of different DNA molecules, but also the preparation of heterogeneous materials containing organic or inorganic material organized by DNA. Examples for heterogeneous complexes are networks of DNA-modified nanoparticles [1, 2] or the DNA-directed self-assembly of proteins [3]. Such complexes are created by the application of DNA-modified units. This approach bases on a connection of DNA to solid surfaces, as needed for the fabrication of DNA-modified nanoparticles [4] or the envisioned preparation of DNA constructions on microstructured chip surfaces [5]. DNA chip technology is another field with growing importance relying on the binding of DNA onto solid surfaces [6]. So the characterization of this binding is of great interest for a range of different applications. Interesting parameters regarding the immobilized DNA-molecules are density, activity, and homogeneity of distribution. Although spectroscopic and optical methods yield information about the density, this information is usually averaged over regions in the upper micrometer or millimeter range. Especially for the field of DNA-based construction, the interesting feature size is often in the lower nanometer range, which is hardly accessible by these methods. Moreover, the activity and/or the variation in surface density in these dimensions cannot be provided. Another set of techniques is based on the use of labeled DNA molecules with a sequence complementary to the immobilized DNA. After incubation with the DNA-modified surface a binding of the labeled DNA is observed. This binding depends on the surface density of immobilized

CP640, *DNA-Based Molecular Construction: International Workshop,* edited by W. Fritzsche
© 2002 American Institute of Physics 0-7354-0095-4/02/$19.00

(complementary) DNA as well as the efficiency of binding as a representation for the activity. Besides this access to the activity of the immobilized DNA, a quantification of the observed signal yields density information. Depending on the used label, the visualization could also give information about the homogeneity. The two standard labels in the biochemical laboratory are radioactive markers and fluorescence dyes. Radioactive labeling is highly sensitive and allows quantification, however, for safety reasons it is only applicable in specialized groups and not easily accessible. Moreover, the lateral resolution is not sufficient for the characterization of binding on microstructured surfaces. Although fluorescence dyes provide a high sensitivity reaching the single molecule level, the lateral resolution is limited to the medium nanometer range due to the optical readout principle. Because the signal of fluorescence dyes depends on the physicochemical environment and is influenced by irreversible processes resulting in the degradation of the dyes (e.g. by bleaching), quantification is rather difficult. Especially for higher surface densities, the location of every single dye molecule cannot be determined.

A new labeling method based on nanoparticle labeling promises to overcome these problems. Nanoparticles, as known from optical and electron microscopy, are long-term stable. Using scanning force microscopy (SFM), the location of each individual particle can be easily determined, allowing a precise qualitative and quantitative characterization of the binding. Based on the visualization, information regarding the density and the distribution are accessible. However, SFM is a rather slow method needing sophisticated (expensive) equipment. The demand for a faster detection (with preferably parallel readout) lead to the application of optical techniques for nanoparticle readout in chip technology [7, 8]. This paper presents first steps toward an application of this technique in molecular nanotechnology for the characterization of surface layers of immobilized DNA.

MATERIALS AND METHODS

Microstructured Substrates

Standard microscope glass slides were separately cleaned with acetone, ethanol and water for 10 min by sonication prior to activation. The chips were immersed into a solution of v H_2O / v H_2O_2 / v HCl (1:1:1) for 10 min, washed with water and dried at 80 °C for 10 min. Then, the chips were incubated in a solution of 1 mM octadecyltrichlorsilane in dried toluene in an argon atmosphere at 40 °C for 4 h. Afterwards the substrates were covered with a photo resist, a pattern of 4 rows, 8 squares of 1 x 1 mm each row was opened in the resist layer exposure and development of the resist. Then the binding areas (4x8 pattern) were opened with oxygen plasma etching and the resist was removed.

DNA Immobilization On The Substrates

The chips were activated again in the solution described as above and then refluxed at 70 °C for 6 h in a solution of 10 mM 3-glycidyloxypropyltrimethoxysilane in dried toluene. After the reaction the chips were washed with toluene (2 x 10 min), ethanol (2 x 10 min), and water (2 x 10 min).

The 5' amino-modified DNA for surface-immobilization was diluted with 0.2 M KOH to yield a 50 μM concentration. 1 μL droplets of that solution were applied on every designated binding spot and the chip was incubated in a humidity chamber at 37° C for 1 h. Afterwards, the chip was washed with a solution of 0.1 % Triton-X100 (1 x 5 min), HCL solution pH 4.0 (2 x 2 min), 0.1 mM KCL solution (1 x 10 min), and H_2O (1 x 1 min). Subsequently, the chip was incubated in a blocking solution with 50 mM ethanolamine, 0.1% SDS in H_2O, pH 9.0 at 50 °C for 30 min, then rinsed with H_2O (1 x 1 min) and dried.

DNA-Gold Nanoparticle Complexes

The 3'-thiol modified 20mer oligodeoxynucleotides (obtained from Bioscience, Jena, Germany) were synthesized using standard phosphoramidit chemistry. The 3'-dodecylthiolated oligonucleotides were desalted by NAP-10 column (Pharmacia) and preincubated with gold nanoparticle solution (30 nm diameter, British Biocell) for 16 h at room temperature in a ratio of 0.33 nM gold and 200 nM DNA. Another incubation was carried out after adjusting the solution to 0.1 M NaCl, 10 mM sodium phosphate buffer, pH 7.0, for 40 h at room temperature. The DNA-nanoparticle complexes were repeatedly washed with buffer and redispersed in 0.1 M NaCl, 10 mM sodium phosphate buffer at pH 7.0.

Solutions with higher concentrations were prepared by centrifugation and redispersion in a corresponding lower volume of buffer.

DNA Sequences

Immobilized DNA
complementary:
H_2N-$(CH_2)_6$ -CCC TAG AAA ATT GAG AGA AGT CCA CCA CGA
non-complementary:
H_2N-$(CH_2)_6$ –TAC ATA CTT ACA TAC TTA CAT ACT TAC ATA

Linker DNA:
GGG ATC TTT TAA CTC TCT TCA GGT GGT GCT- CAG AT- CTG AGA CAC CAT AAC ACT CC

Label DNA (complementary to linker DNA):
GAC TCT GTG GTA TTG TGA GG-$(CH_2)_{12}$ – S -nanoparticle

Sandwich Hybridization

The linker DNA were diluted to 10 μM in 2 x SSC. 1 μL of that solution was then applied on every immobilized probe. Hybridization was done in a humid chamber at 65 °C for 4 h. Not hybridized and non-specific probes were removed by washing with PBS containing 0.1 % Tween (3 x 5 min).

The adopted solution of DNA-nanoparticle complexes had an optical density of 2.0 OD at 525 nm. Subsequently, that solution was applied on the chip and hybridization was carried out in a humidity chamber at 46 °C for another 4 h and cooled slowly to 20 °C. In a different concentration experiment, the solution of DNA-nanoparticle complexes was diluted to 2.0, 1.0, 0.5, and 0.25 OD, respectively. The chip was then washed with 2 x SSC containing 0.2% SDS for 10 min, 2 x SSC for 10 min, 0.2 x SSC for 10 min.

Silver Enhancement And Detection

The chip was immersed into a silver enhancer solution, which was composed of silver acetate (80 mg in 40 mL H_2O) and hydroquinone (200 mg in 40 mL citrate buffer, pH 3.8). The silver enhancer solution was then placed in a dark box for 10 min. After the treatment of the chip with the silver enhancer solution, the chip was rinsed with water, and then air-dried.

The chip was finally imaged using a flatbed scanner AGFA Duoscan T 2500 in reflective mode.

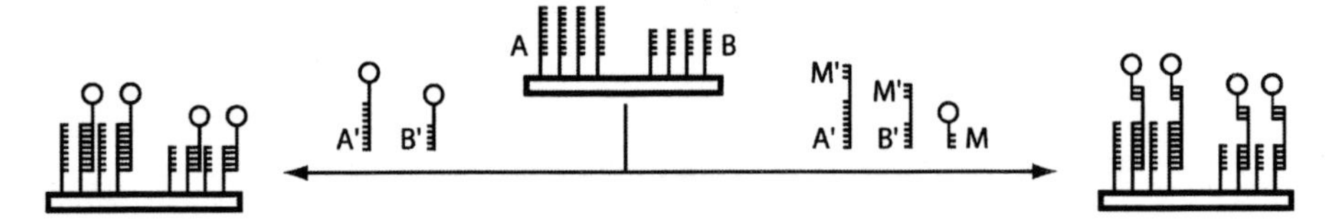

FIGURE 1. Scheme of nanoparticle-labeling for the detection of surface-immobilized DNA. Left: Direct labeling; Right: Indirect (sandwich) approach using a linker molecule.

RESULTS AND DISCUSSIONS

Nanoparticle Labeling

The general scheme for the characterization of surface-immobilized DNA using nanoparticle labels is shown in Fig. 1. A substrate holds two areas each modified with a different DNA molecule containing the sequences A and B, respectively (center). A direct approach for labeling is shown in the left part: DNA complementary to A and B labeled with nanoparticles is used for detection, based on hybridization of the labeled DNA at surface regions containing complementary DNA prior to a readout of the labels. A more indirect approach using linker DNA molecules which are complementary to the molecules of interest (A or B) but also hold another sequence which binds to a nanoparticle-labled DNA (M) is described in the right part. This approach resembles the sandwich setup often applied in immuno-diagnostics. It requires additional molecules and is more difficult to perform due to its two-step hybridization procedure. However, it minimizes the number of required kinds of nanoparticle-labeled molecule. Although the preparation of DNA-modified nanoparticles is usually based on a standard thiol-gold chemistry, it is still no standard procedure and the characterization of the product is difficult. On the other side, the design and preparation of the linker DNA is easily accomplished, and the successful implementation of nanoparticle labeling for a comparable (indirect) setup was demonstrated [9]. So the application of the sandwich approach is preferable in the presence of a variety of sequences, as it will be the case in the further development of a DNA-based molecular nanotechnology.

Readout Schemes

There are various methods for the detection of nanoparticle labels on a solid surface, including scanning force microscopy (SFM), optical techniques and electrical schemes.

SFM is a microscopical method yielding the three-dimensional surface topography with a resolution in the lower nanometer range. It allows the resolution of individual nanoparticles, thereby enabling the detection of single binding events (Fig. 2c). As a serial method, SFM imaging is slow and yields only a limited field of view (below 150 µm scan size) of the overall surface. So the characterization of larger areas is restricted, especially in the case of unknown homogeneity of the nanoparticle distribution.

The recently introduced electrical detections scheme [10,11] has the potential of higher throughput, but its need for prestructured microelectrodes excludes it as a standard method in molecular nanotechnology. However, for a limited field of experiments with included electrode structures, it could be a valuable addition.

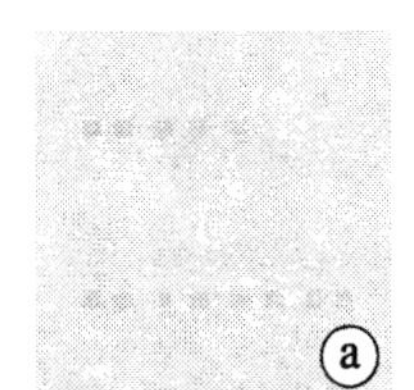

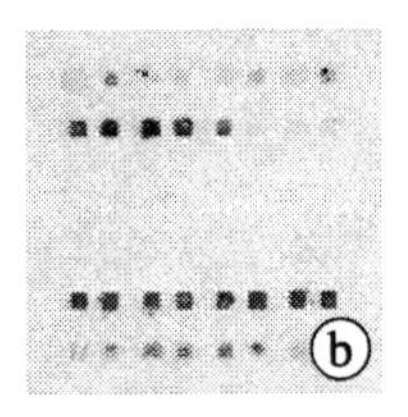

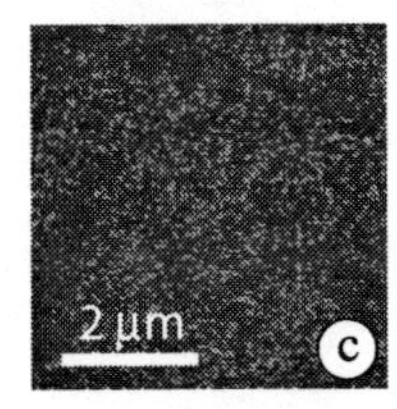

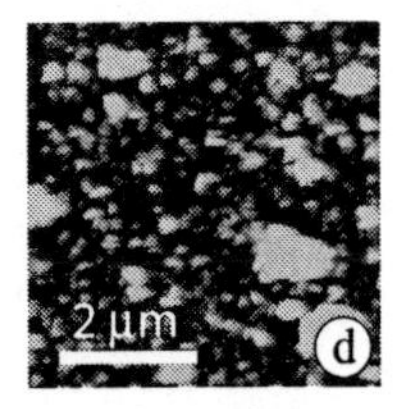

FIGURE 2. Metal enhancement of nanoparticle labeling. a, b) Optical contrast (reflection) of micro-structured squares of 1x1 mm which were modified with DNA and incubated with complementary DNA labeled with gold nanoparticles before (a) and after (b) silver enhancement. c,d) Scanning force microscopy of surfaces with specifically bound nanoparticles before (c) and after (d) enhancment.

Optical techniques promise the detection of nanoparticle-binding directly [8] or after an enhancement step [7]. They require only standard equipment as e.g. microscopes or scanners, and provide the needed throughput by their parallel character. Areas of 1x1 mm modified with DNA patterned on a glass slide (cf. Fig. 2a,b; 3a; 4a) were used for the experiments described here. They contained different

DNA sequences, which were probed with nanoparticle-labeled DNA. After incubation with labeled DNA and subsequent washing procedures to minimize unspecific binding, the DNA-modified surface regions revealed immobilized nanoparticles. Depending on the surface density, it can be detected by the naked eye or a standard flatbed scanner, or it is only resolvable by SFM in the case of low surface density of nanoparticles (Fig. 2c). The latter case points to the missing sensitivity as a serious limitation of the optical detection. A low sensitivity restricts the dynamic range, as it was visible in experiments with different concentrations [8], thereby hampering a broad applicability of the readout principle. So the optical readout of gold nanoparticles alone is not sufficient for a comprehensive characterization of DNA immobilized on surfaces (Fig. 2a), a signal enhancement is required.

Enhancement Procedures

The needed improvement in signal level can be achieved using specific metal deposition on the immobilized nanoparticles. This technique was adapted from light and electron microscopy [12] and already applied for signal enhancement of nanoparticles in a DNA chip-like application [7, 13]. Its application in molecular nanotechnology is based on two advantages: The increased sensitivity (thereby extending the dynamic range) and the potential to visualize small numbers or even single particles using optical readout. The latter point is subject of ongoing investigations [14]. The increased sensitivity is demonstrated in Fig. 2b, showing the nanoparticle-labeled sample from Fig. 2a after a silver enhancement procedure: The signal is significantly increased and easily detectable. So instead of a scanner with its high price tag or a microscope with its limited field of view, standard flatbed scanners can be used to streamline the detection at significantly reduced costs [7, 13].

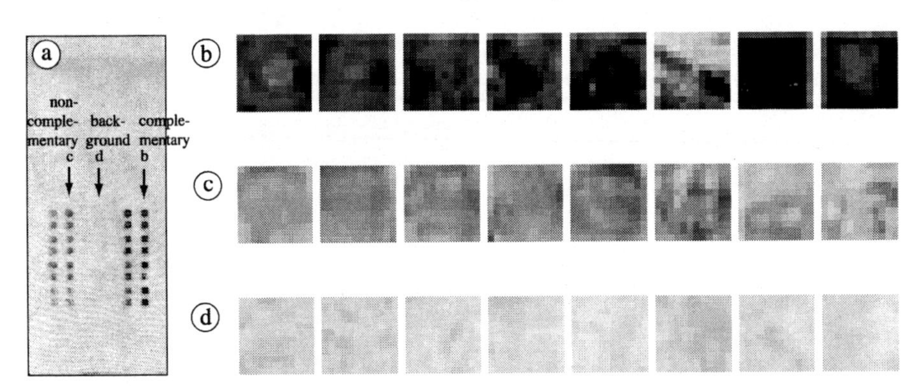

FIGURE 3. Demonstration of the specificity of nanoparticle labeling. a) Overview of a slide containing 4 columns of 1x1 mm squares modified with DNA after incubation with nanoparticle-labeled DNA and silver enhancement. b-d) Zoom of the squares. b) Specific binding results in a strong signal. c) Non-complementary DNA leads to a low signal, probably due to unspecific binding. d) The background signal due to unspecific silver staining is negligible.

Specificity

An important point for this envisioned use of optical nanoparticle detection for molecular nanotechnology is the specificity of the reaction of labeled DNA with the surface-immobilized DNA. The specificity of hybridization of nanoparticle DNA to surface-immobilized DNA was demonstrated in the case of gold [15] or silicon oxide [9] substrates using SFM visualization. To confirm these results for our set of parameters, experiments were conducted using surface-immobilized sequences both complementary and non-complementary to the applied nanoparticle-labeled DNA. Fig. 3a shows an overview of the used slide with four columns of microstructured binding areas, after incubation with labeled DNA and the silver enhancement. A fifth column in the center region (which was not previously microstructured) was also investigated and served as control. This control showed that the silver enhancement is highly specific, and shows hardly any signal outside of the microstructured areas in the four columns (see the zoom in Fig. 3d). The right column represents the complementary sequence, and exhibits the strongest signal (zoom of the spots in Fig. 3b). Taking the low unspecific silver binding into account, this signal points to a strong binding of nanoparticles to the spots in this row. Unspecific binding onto immobilized DNA would be another possibility. A significant contribution of this effect could be ruled out by testing non-complementary DNA. This was accomplished in the second row; the results are visualized in Fig. 3c. Although a weak signal is visible in comparison of the unlabeled background (cf. Fig. 3d), it is significantly below the specific signals form Fig. 3b. These experiments demonstrated the large contribution of specific binding to the observed nanoparticle binding with a subsequent enhancement step.

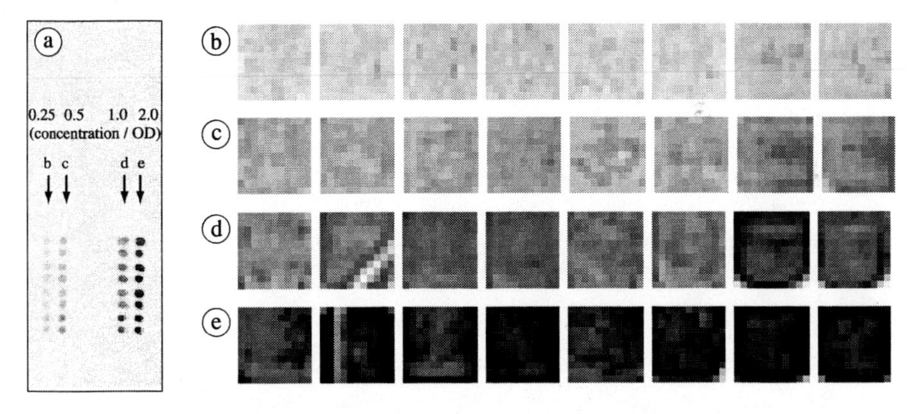

FIGURE 4. Optical signal in dependence on the concentration of label DNA. a) Overview after nanoparticle labeling and silver enhancement showing the different concentrations. b-d) Zoomed images of the microstructured squares, a raise in signal becomes apparent with increasing label DNA concentration.

Quantification

Beside the demonstrated specificity, the potential for quantification is the next important parameter. The detected gray value has to be correlated to a surface density of nanoparticles and subsequently to the number of immobilized DNA molecules in a given area. The first step towards this aim is the study of signal-response relationship for a given concentrations of nanoparticles and varying concentrations of immobilized DNA. An overview of a slide with DNA-modified areas which were incubated with different concentrations of label DNA is shown in Fig. 4a, the image was taken after silver enhancement. The enlarged areas with immobilized DNA in Fig. 4b-e illustrate a step-wise increase in signal from b to e. This enhancement represents an increase in concentration of label DNA. Further studies will elucidate the character of correlation between these parameters. Therefore, the gray values will be extracted from the images and used for data processing.

Study Of Homogeneity

The quality of DNA-modified binding areas for applications in molecular nanotechnology is mainly influenced by the surface density and the homogeneity. Whereas the density was discussed in the previous paragraph, the homogeneity of the distribution of active groups is crucial for the efficiency of further binding reactions. SFM yields the distribution of nanoparticle labels in a limited area in the sub-100 μm range (Fig. 2c). Larger areas could be studied by optical means after silver enhancement, if the distribution of the optical signal correlates with the original distribution of the nanoparticles. The enhancement itself is highly specific and homogeneous, especially for a sub-monolayer state, as a microscopic study of the growth of nanoparticles showed [11]. A typical cause of inhomogeneous surface immobilization is droplet incubation. This effect is known from DNA chip technology, where microarrays are produced by spotting droplets of DNA onto a substrate. It often results in ring- or doughnut-shaped intensity distribution, probably induced by drying artifacts in combination with a concentration gradient in a droplet on the substrate. The bright bottom corners of the right squares in Fig. 4d and e point to such a phenomenon. Another cause for areas without modification are scratches, which are often due to manipulation of the substrates. A typical example is visible in the second square in Fig. 4d. Also the square below it points to such an artifact.

CONCLUSIONS

These examples demonstrate the potential of enhanced nanoparticle-labeling for the characterization of DNA-modified surfaces. It allows the detection of the presence of the molecular layer, beside a potential for quantification and the investigation of properties like distribution and homogeneity. Further work will reveal the signal-response behavior for quantification, and aim at the establishment of an ultrasensitive optical detection approaching the ultimate single particle limit.

ACKNOWLEDGMENTS

We thank J. M. Köhler for his initial contributions to the nanoparticle research in our group, K. Kandera and D. Horn for assistance with microstructuring of the substrates.

This work was supported by the DAAD (scholarship to G. -J. Z.) and the DFG (FR 1348/3-4).

REFERENCES

1. P. Alivisatos, K. P. Johnsson, X. Peng, T. E. Wilson, C. J. Loweth, M. P. Bruchez Jr., and P. G. Schultz, Nature **382**, 609 (1996).
2. Mirkin, R. L. Letsinger, R. C. Mucic, and J. J. Storhoff, Nature **382**, 607 (1996).
3. M. Niemeyer, B. Ceyhan, S. Gao, L. Chi, S. Peschel, and U. Simon, Colloid and Polymers Sciences **279**, 68 (2001).
4. J. J. Storhoff, R. Elghanian, R. C. Mucic, C. A. Mikron, and R. L. Letsinger, Journal of the American Chemical Society **120**, 1959 (1998).
5. Csaki, R. Möller, and W. Fritzsche, Expert Review in Molecular Diagnostics **2**, 187 (2002).
6. M. J. Heller, Annual Reviews of Biomedial Engineering **4** (2002).
7. T. A. Taton, C. A. Mirkin, and R. L. Letsinger, Science **289**, 1757 (2000).
8. J. Reichert, A. Csáki, J. M. Köhler, and W. Fritzsche, Analytical Chemistry **72**, 6025 (2000).
9. R. Möller, A. Csáki, J. M. Köhler, and W. Fritzsche, Nucleic Acids Research **28**, e91 (2000).
10. S. J. Park, T. A. Taton, and C. A. Mirkin, Science **295**, 1503 (2002).
11. R. Möller, A. Csáki, J. M. Köhler, and W. Fritzsche, Langmuir **17**, 5426 (2001).
12. G. W. Hacker, in *Colloidal Gold: Principles, Methods, and Applications*, edited by M. A. Hayat (Academic Press, 1989), Vol. 1, p. 297.
13. W. Fritzsche, A. Csaki, and R. Möller, SPIE **4626**, 17 (2002).
14. Csáki, P. Kaplanek, R. Möller, J. M. Köhler, and W. Fritzsche, in preparation (2002).
15. Csáki, R. Möller, W. Straube, J. M. Köhler, and W. Fritzsche, Nucleic Acids Research **29**, e81 (2001).

DNA On Surfaces: Adsorption, Equilibration And Recognition Processes From A Microscopist's View

Giampaolo Zuccheri,[a] Anna Bergia,[a] Anita Scipioni,[b] Pasquale De Santis,[b] Bruno Samorì[a]

[a]*Department of Biochemistry, University of Bologna and INFM, via Irnerio, 48 – 40126, Bologna, Italy*
[b]*Department of Chemistry, University of Rome La Sapienza, Piazzale Aldo Moro,5 – 00185 Rome, Italy*

Abstract. The availability of a number of methods to controllably adsorb DNA on solid surfaces is useful to researchers working in different fields, such as structural biology, biophysical chemistry, diagnostics, sensorics, and nanotechnology. In this paper, we review some of the methods that have been devised in the last years to solve this problem. Thanks to the Scanning Force Microscope, we have recently been able to study the dynamics of DNA molecules adsorbed on mica. From the statistical analysis of the shapes of properly designed DNA molecules, we have also discovered an unexpected base-sequence specificity in the adsorption of intrinsically curved DNA molecules on mica.

The recent developments and widespread use of DNA chip technology and DNA-based sensorics has greatly fostered the interest in controlled DNA attachment to surfaces. In the last years, many protocols for covalent attachment of oligonucleotides or long ds-DNA molecules on metallic or non-metallic surfaces have been reported and we can expect that in the near future many more will be offered to researchers. Many efforts are also dedicated to the controlled anchoring of nucleic acids at surfaces with nanometer precision [1].

Most applications of DNA and DNA-based molecular constructions to nanotechnology will involve the deposition of DNA on surfaces, either as part of the assembly and functioning of the constructions [2] or as a method to check the success of complex building procedures [3]. The adsorption of DNA at very flat surfaces has also been a primary concern for researchers using the Scanning Force Microscope (SFM) to study the structure and function of nucleic acids [4].

This article is focused on the topic of DNA adsorption and on the SFM study of the conformation and dynamics of adsorbed DNA molecules. At the beginning of the 90's, all the efforts concentrated on the reproducibility and the efficiency of the methods. A stable adsorption is necessary to prevent the scanning probe from sweeping the molecules around on the surface. In the last few years, the attention has centered on the modulation of the adsorption strength and on the specificity of the adhesion of DNA. Of concern is also the possibility of using the mildest conditions, so that fragile complexes or constructions are not altered or disrupted during the deposition on the surface.

CP640, *DNA-Based Molecular Construction: International Workshop*, edited by W. Fritzsche
© 2002 American Institute of Physics 0-7354-0095-4/02/$19.00

Very recently, our group discovered the unexpected capability of an inorganic surface to recognize DNA sequences and superstructures, by differentially adsorbing them. Much effort is nowadays paid to the use of DNA molecules for the building of self-assembling nano-structures [5]. So far this has been done by using the self-assembling information that the sequence of these molecules contains. A higher level of complexity could be reached when the same DNA molecules or nanostructured molecular constructions based on them can be recognized and adsorbed by a crystal surface.

METHODS FOR THE ADSORPTION OF DNA AT SURFACES

In this section, without any claim of completeness, we will briefly review some protocols that have been used to drive the non-covalent attachment of DNA at surfaces, especially in the context of the SFM experiments. We chose not to dwell on the plethora of methods developed to covalently anchor DNA at surfaces.

Electrochemical adsorption of DNA

One of the first methods employed to adsorb DNA on a flat surface was discharging DNA onto a flat gold electrode [6]. The development of Scanning Tunneling Microscopy (STM) had brought methodologies to prepare flat surfaces of gold (such as flat facelets on a small gold ball). This deposition method can even be reversible, nevertheless it did not encounter much success since it requires a complex instrumental set-up and it produces data whose biological relevance could be questioned.

Dehydration of solutions of DNA on mica

The most straightforward method to spread DNA on a surface would be to dry a dilute solution of molecules on a suitable substrate. To promote the adsorption of polyelectrolytes, the substrate must be highly hydrophilic. As for any other substrate that could be used for SFM analysis, the deposition surface must be very flat and should not provide any structural features that could be mistaken for the specimen: ideally, it should be featureless and very clean. The required specifications led to muscovite mica, already in use amongst electron microscopists, which soon became the most frequently used substrate for the adsorption of biological macromolecules for the SFM. Mica can be easily cleaved to provide the microscopists with atomically flat and highly hydrophilic surfaces.

The choice of the solution conditions for deposition by dehydration is crucial. While the pH must be buffered, to ensure the chemical stability of the DNA and the biological relevance of its structure, the chemicals used for the buffering represent a serious problem at the moment of dehydration, since they would crystallize and precipitate on the surface. One of the methods employed was to dissolve the DNA in ultrapure deionized water, making sure that the pH would not be too low, and to dehydrate a drop of such a solution on freshly cleaved mica [7]. The drawbacks of such a protocol were the scarce efficiency of the adsorption, and the fear that DNA would assume undesired conformations in the extremely low ionic strength medium. A clever trick was the use of ammonium acetate to buffer the DNA solution [8-10]. This salt is volatile and it should evaporate completely with the water, leaving only the DNA on the surface. The method is reported as very efficient (it can use DNA solutions as dilute as 0.3 μg/ml) but it does not allow a fine control of the ionic strength of the solution as the water evaporates. Alternatively, some researchers dried specimens that contained non-volatile buffers (such as Tris), and sometimes resorted to rinse the spreads after dehydration [11].

A very high electrolyte concentration near the surface could explain the very condensed DNA shapes sometimes shown by the SFM images obtained with this method.

The control of the DNA concentration is critical: in low ionic strength solutions (and presumably also with the aid of the hydrophilic mica surface), extended portions of the DNA chain can melt. At the moment of dehydration, complex networks can result from the intramolecular and intermolecular hybridization of neighboring chains. These frameworks can grow up to completely fill the mica surface with DNA networks of unpredictable shape and inhomogeneous local structure.

Silane treatment of mica

Yuri Lyubchenko and his collaborators found a method to adsorb DNA onto mica by treating the surface with amino-propyl triethoxy-silane (APTES) [12,13]. The free surface-attached amino group is sometimes methylated after silane binding to mica [13]. In water, the positively charged ammonium can bind the negatively charged DNA (and RNA) backbone and the adsorption is strong enough for SFM imaging. There are several available protocols for surface functionalization with APTES: the freshly cleaved mica can be exposed to APTES vapors [12,13], or to APTES solutions made in acetone [14], toluene, dimethylformamide [13] and water [15]. Sometimes a partially hydrolyzed silane solutions is preferred [16]. Such a derivatized surface is referred to as AP-mica.

AP-mica adsorbs DNA very efficiently (down to concentrations of 0.01 μg/ml for lambda DNA)[15,17,18] and under a variety of conditions (from 0 to 60 °C and under different ionic strengths). In principle, it should be possible to modulate the strength of adsorption by reducing the density of the aminopropyl groups on mica, but the experiments in the literature all report very strongly attached DNA. As it will be discussed below, one of the main concerns is the control and the modulation of adhesion. Too strong an adhesion that cannot be modulated will prevent surface equilibration of the DNA molecules [19]. It must also be pointed out that most surface treatment operations often produce "dirty" surfaces for not all the experimental parameters can be under strict control. APTES treatment is no exception, and it happens that impurities in the reactants, partial polymerization or inhomogeneous treatments can make a discontinuous surface, where only portions of the surface area are flat enough to be used for SFM analysis [13,16].

One of the concerns in using surfaces treated with APTES is its evidenced ability to condense DNA [13]. Plasmids adsorbed from low salt solutions can form rods and condensed amorphous structures (our unpublished data and [16]). It has been proposed that the DNA molecules could recruit mobile adsorbed silane molecules, which then cause them to condense: heat or vacuum treatments have proved useful in eliminating DNA condensation on treated silicon substrates.

Soluble cations promote the adsorption on mica

The most reliable methods to adsorb DNA on mica imply the use of soluble divalent cations [15,20-23]. Substituting the monovalent cations of mica with divalent cations from a solution (usually magnesium) makes the surface more positive and "activated" toward the adsorption of DNA molecules from a drop of solution laid on it. Glow discharge was also used to make the magnesium-substituted surface more hydrophilic, even though it was seen to be unnecessary [24]. Trivalent cation pretreatment of mica has been shown to work too [25]. The magnesium pretreatment is not necessarily a distinct step from DNA adsorption [21,26], and solutions containing both the DNA and the magnesium salt can be deposited on freshly cleaved mica: this is currently the method of choice.

As a rule of thumb, multivalent cations in solution promote the adsorption of DNA to mica, while monovalent cations decrease it [27]. As studied experimentally by Rivetti and co-workers, in magnesium-mediated DNA adsorption, diffusion transports the DNA to the surface where it is then irreversibly adsorbed. If conditions are properly tuned, the adsorbed molecules still maintain two-dimensional diffusional freedom on the surface up to the moment the layer of water on the surface is dried [19]. Hansma and Laney studied the effect of the type and concentration of the cation on DNA deposition [28].

Even though more efficient in depositing DNA on mica, certain metal cations could affect the structure of the nucleic acids by substituting the counterions they physiologically bear, and could be poisonous for proteins. It has been shown that, quite interestingly, Zn ions cause kinking in DNA [29,30]; furthermore, Zn(II) is poorly soluble in water. Divalent cations can alter the secondary structure of DNA, and they have been seen to affect the contour length of DNA molecules [16], probably by altering the winding of the DNA helix [31]. The optimization of protocols for depositing DNA on untreated mica from solutions containing inorganic cations has made the observations of DNA a routine procedure. If a researcher has a purified sample of DNA, he/she can certainly deposit it on untreated mica and produce high quality images by following these protocols.

Other Charged Molecules In Solution Can Promote DNA Adsorption

Inorganic cations are not the only effective electrostatic means for adsorbing DNA on mica: more complex organic charged molecules also proved effective. Quaternary ammonium salts (benzyldimethylalkylammonium chloride) have been successfully employed to spread DNA for SFM observations [32,33]. The concentration of the detergent used in the reported experiments does not seem to affect protein-DNA interactions.

A recent paper [34] shows that soluble silanes bearing two or three amino groups can promote the adsorption of a water solution of DNA to freshly cleaved mica. The proposed method seems interesting, but the ability of the silanes to condense the DNA must be kept under control.

THERMODYNAMIC EQUILIBRATION OF DNA ON MICA

As reported by Rivetti and co-workers [19] many protocols that produce efficient DNA deposition on surfaces can also lead to kinetic trapping of the molecules on the mica substrate. On the surface, such trapped molecules display a global conformation that should represent the projection of the shape they had in solution: this depends on the path they followed for landing on the surface. On the other hand, if the conditions are properly adjusted, adsorption does not hinder two-dimensional diffusion on the surface, so molecules can equilibrate thermodynamically and assume global conformations that are directed by their physico-chemical nature (that can be studied from their imaging). Many protocols for the pretreatment of mica or the direct deposition of DNA molecules can lead to a strong adsorption and trapping and should be avoided. If the deposition protocol only exploits the ionic exchange properties of mica, it seems enough to always keep a millimolar concentration of Mg(II) in solution to avoid trapping. Other treatments that significantly alter the nature of mica or establish covalent bonds between the mica surface and the DNA molecules must be evaluated on an individual basis, analyzing the statistical global shape of the adsorbed molecules, for example with the methods described by Rivetti and co-workers [19].

If the structural features of interest in DNA are more local, or are studied with methods that analyze local spatial parameters of the adsorbed DNA molecules, then the issue of surface equilibration is expected to be less serious. Many fast local dynamics must compose to change the global shape of a molecule. The local structural features (like bend angles or chain dynamics) could be faithfully responding to the molecular properties even if the global shape of the molecule is not. A deposition protocol that binds the molecules strongly could provide thermodynamically meaningful data only on a smaller scale than others or for shorter DNA molecules: for instance, long DNA molecules diffuse considerably less on the surface of AP-mica than shorter molecules, on a comparable time-scale [17,35]. To harvest the greatest amount of information from SFM experiments, researchers should employ deposition methods that can be tailored to their experiments and to the other boundary conditions.

CONFORMATIONAL FLUCTUATIONS OF INDIVIDUAL DNA MOLECULES DEPOSITED ON MICA

The Time-Lapse Study Of DNA Chain Conformational Fluctuations

By adjusting the ionic strength and the concentration of divalent and monovalent ions, as supra described, one can modulate continuously and reversibly the strength of adhesion of the molecules and so modulate the DNA mobility on mica. This allowed us to observe the conformational fluctuations of single DNA molecules in real time [36]. For SFM studies, the molecules cannot move too fast, otherwise the scanning probe will not be able to image them. The conformational space thermally accessible to the fluctuations of a molecule is shaped by the local persistent curvatures and helix deformability of its chain, one can therefore learn about these two structural and mechanical properties from a sequence of images showing gradual changes of the shape of the molecule due to its thermal fluctuations.

We decided to study the dynamics of single DNA molecules in order to establish a method to map the sequence-dependent nanoscale conformational and mechanical properties along a DNA chain (A. Scipioni et al., manuscript submitted for publication). We have deposited several linear DNA molecules on the surface of freshly cleaved mica and imaged them in fluid, without ever dehydrating the specimen. We could image the molecules for sufficiently long periods (up to 4 hours), so that it could be possible to make movies out of the molecule motions, and to quantitatively analyze the time-dependence of the structural parameters (Fig. 1).

Superimposing all the profiles recorded during this observation (Fig. 2) one can notice that the global shape of the molecules does not change significantly: the conformational space that the molecules spanned during this time was a very limited portion of that thermally accessible to their fluctuations, due to the relatively strong adhesion to the surface that seriously slows down the long-range chain dynamics.

The ensemble of the global molecule profiles did not equilibrate over the time of the experiment: in spite of that, the local dynamics turned out to be ergodic, as quantitatively determined by a recently-developed method that makes it possible to evaluate the scale-length of the chain equilibration. This method is based on the comparison of the local time-averaged curvature modulus and its standard deviation (A. Scipioni et al., manuscript submitted for publication).

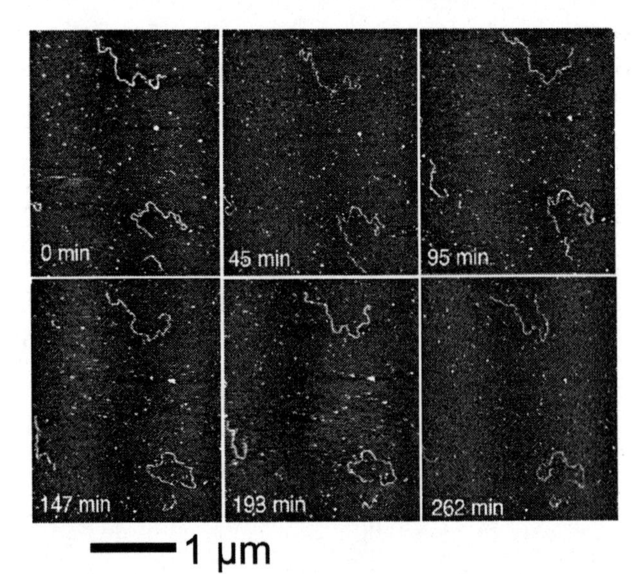

Figure 1. Sequence of SFM images recorded in an aqueous buffer of two linearized pBR322 DNA molecules. The overall time of the observation was 4 hours. Each molecule is 1.5 μm long.

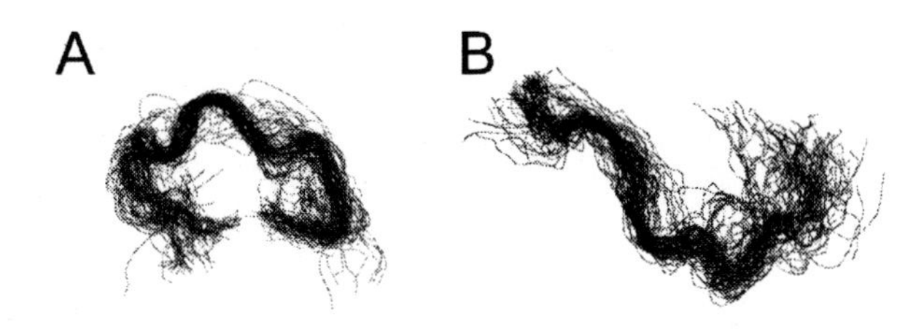

Figure 2. Superimposition of 107 and 94 traces derived from SFM images of the two linearized pBR322 DNA molecules shown in Figure 1. The individual images were obtained at time intervals of about 2 min.

The size-scale where this fast equilibration takes place is of the order of a few tens of nanometers, very close to that of the interaction between proteins and DNA. Our experiments represent a direct demonstration that even in conditions that inhibit free global motions of long DNA chains, the local interactions between DNA and other molecules can still be governed by equilibrium thermodynamics. The importance of this finding might be broadcast: DNA molecules in living organisms are contemporarily involved in many complexes with bulky molecules and membranes that determine a significant hindrance to the motion of large portions of the very long genomic DNA molecules, still the single local interactions can take place under thermodynamically equilibrated conditions. In DNA nanotechnology, the nanostructures are frequently built or designed to function on a solid surface; even though attached in a relatively strong manner, the ergodicity of the small chain dynamics of DNA could imply that the local interactions would still take place freely and in a reproducible and predictable fashion.

Population Averaging From Single-Molecule Microscopy Data

An equilibration process that takes place only locally is not sufficient for studying the shapes of DNA molecules exhaustively. To this end, a complete equilibrium ensemble of chain profiles is needed. According to the ergodic principle, the ensemble of the profiles assumed over time by only one molecule is statistically identical to the ensemble of profiles of a population of same molecules at a given time. The time of the population sampling might be that of the dehydration of a DNA specimen on a surface, when the movements of a very large number of molecules are halted. In experimental terms, a study involving a population average on a dehydrated specimen is incomparably easier than one requiring a time average over a few molecules observed for a long time in solution.

Dealing with ensembles of profiles of different molecules implies the loss of the total sample homogeneity of a single-molecule experiment. One of the immediate corollaries is that the orientation of the base sequence in a molecule is not the same for all instances of the population in a microscope image, since each non-symmetric molecule can be "read" beginning from one end or the opposite. A classical solution to this problem has been the labelling of one end with a bulky protein [37]. In doubt that the protein could modify the mechanical properties of the proximal sections of the chain, or alter the surface adsorption properties of that chain section, we took an alternate route: we prepared a palindromic dimer of the molecule under study [38]. In palindromic molecules, the sequence is the same reading from either end. The use of a palindrome introduces a binary symmetry that allows the univocal correspondence of a position along the DNA trace in the microscope images with a position in the base sequence. This symmetry is also reflected in the stereochemistry of the molecules and therefore in all the average structural and physical properties experimentally detected along the DNA chain. The presence of two symmetrically arranged identical curved sections acts also as an internal gauge of the statistical completeness of the experimental data set, since a relevant symmetry of the properties comes out in the experimental results once a minimum sufficient population size has been reached. This symmetry allows also a doubling of the curvature data obtained from the analysis of the conformation of one molecule, since each sequence motif is present twice per molecule.

We prepared two palindromic dimers using a DNA tract of the pBR322 plasmid containing one markedly curved segment, joining two identical monomers either in a head-head or in a tail-tail arrangement (the tail-tail is the one with the curvatures near the ends, while the head-head is the one with the curvatures near the center of the molecule, see Fig. 3). From more than a thousand molecular profiles or each dimer (Fig. 3), we calculated the population-averaged local chain curvature as a function of the position along the chain.

To study the shapes of the molecules, we computed the local vector curvature along the chain, from the angular deflections of the chain axis. The vector curvature takes into account also the direction of chain curvature (truly, only their sign in 2D). The experimental curvature plot of the head-head dimer shows two marked curvatures near the center, in good agreement with the theoretical predictions for that base sequence [38] (Fig. 4B). Because of the inversion of the sequence of the two halves, the two curved tracts are predicted to be near the ends in the tail-tail dimer (Fig. 4B). Again, the theoretical and the experimental curvature plots are in good agreement. This result has been the first clear-cut evidence that from SFM images of a population of DNA molecules it is possible to map the local conformational properties along the chain [38]. Quite interestingly, along with the local chain curvature, also the local chain flexibility can be evaluated. Contrary to some expectations, our data seem to show that sections of high average curvature are also more flexible than the others.

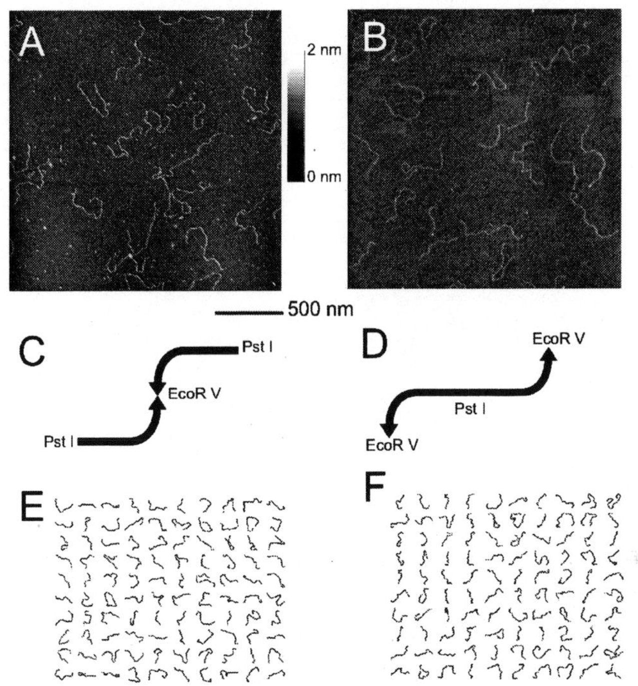

Figure 3. A and B) SFM images of the head-head and tail-tail palindromic dimers (respectively) deposited on freshly cleaved mica and imaged in air. C and D) Schematic representation of the head-head and tail-tail DNA palindromic dimers obtained by joining two Eco RV-Pst I fragments of pBR322. E and F) A sample of the profiles of the dimers obtained from the digitalization of the SFM images (head-head dimers in E, tail-tail in F). The thermal energy causes a high variability in the shapes the molecules assume on the surface: the statistical analysis of the molecular shapes allows us to study the structural properties of the two dimers.

RECOGNITION OF THE DNA SEQUENCE BY AN INORGANIC SURFACE

As evident in Fig. 4, not only the location of the curvatures in the plots changes from one dimer to the other, as expected from the half-sequences inversions, but also their signs are flipped. This is an evidence that the average S-shaped molecules (Fig. 3C and D) that are built by assembling two curved monomers have the tendency to adsorb on the surface always with the same face of the monomers: as it can be seen in Fig. 4C and D, one average experimental shape is obtained from the other by rotating the monomers about an axis perpendicular to the page, not by flipping them, so that they always expose to the substrate the same face. This surface preference effect is not very strong for the pBR322 dimers: in order to confirm this result and to find out more about it, we prepared DNA constructs made of DNA with stronger curvatures.

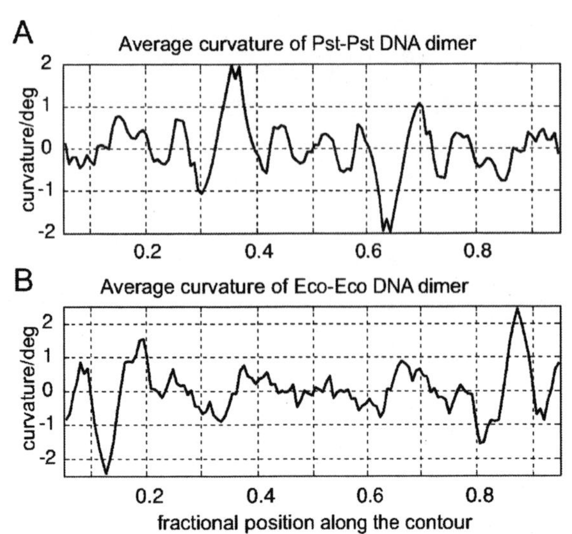

Figure 4. Diagrams of the sequence-dependent curvature of the two DNA palindromic dimers derived form the head-head or tail-tail dimerization of the EcoR V-Pst I fragment of pBR322. (A) Symmetrized averaged experimental curvature profile for more than a thousand head-head dimers. (B) Symmetrized averaged experimental curvature profile for more than a thousand tail-tail dimers. The symmetrization improves the statistical averaging by superimposing the data from the two halves of each molecule.

Extended intrinsic curvatures along a DNA chain are often originated by a periodical recurrence of A-tracts phased with the B-DNA periodicity, i.e. tracts of 4 to 6 adenine steps centered approximately every 10.5 base pairs [39]. A tract of DNA with very extensive and almost perfect A-tract phasing is readily available in the Kinetoplast DNA of the Trypanosomatidae Protozoan *Crithidia fasciculata* [40]. 16 phased A-tracts (out of the 18 present) make this 211 bp DNA segment the most highly-curved natural DNA presently known. From theoretical evaluations it can be seen that such a recurrence of A-tracts on a DNA strand determines a strong curvature and, at the same time, a peculiar spatial segregation of the A bases [41,42]. Due to the extensive phasing and to the structural origin of the DNA curvature, all phased A bases are located, on the average, on one side of the molecular plane, while T bases are, consequently, on the opposite side. This peculiar base segregation, inherent in this choice of DNA sequence, is instrumental to finding out about the importance of base composition in the just discovered phenomenon of preferential DNA adsorption.

We prepared two palindromic dimers from an about 500 bp DNA segment containing the 211 bp *Crithidia fasciculata* DNA. The two approximately 1 kbp DNA palindromes were the result of either a head-head or tail-tail dimerization of the *Crithidia* segment, in a similar fashion as we did for the pBR322 segments. We collected a large pool of SFM images (about 1500) of both palindromic DNA constructs deposited on the surface of freshly cleaved muscovite mica. The molecular shapes were extracted from the SFM images and analyzed as in the previous case, yielding the curvature profiles reported in Fig. 5.

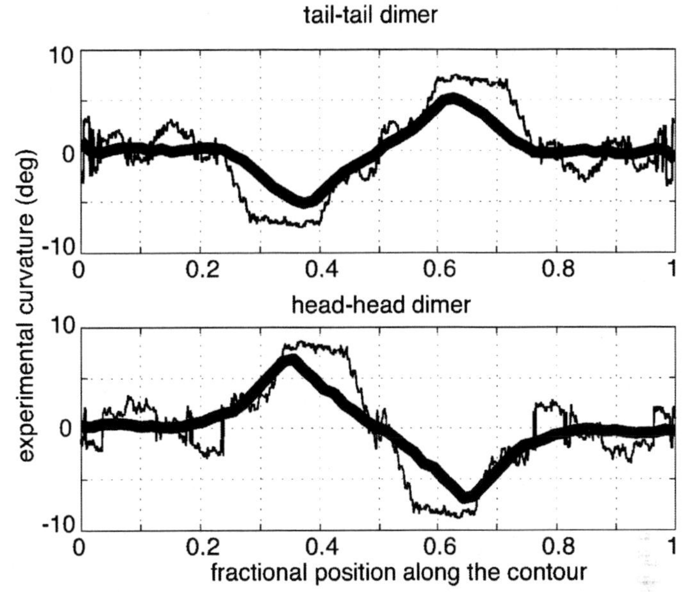

Figure 5. Diagrams of the sequence-dependent curvature of the two DNA palindromic dimers derived form the dimerization head-to-head or tail-to-tail of a *Crithidia fasciculata* DNA fragment. A) Symmetrized averaged experimental curvature profile (thick line) for more than a thousand tail-to-tail dimers and theoretical curvature profile (thin line). The theoretical curvature profile is plotted off-scale to highlight the similarity of the trends of the two profiles. B) Averaged experimental curvature profile (thick line) for more than 1000 head-head dimers and theoretical curvature profile (thin line). The theoretical curvature profile is plotted off-scale.

The sigmoidal shape of the population-averaged curvature plots in Fig. 4 was confirmed also for this DNA dimers but the effect was much stronger than with the less curved sequences. The two oppositely signed profiles revealed that the two palindromic dimers were on the average deposited with two shapes that were mirror images one of the other.

The palindrome dimer termed 'head-head' displayed a positive curvature for the first half and a negative curvature for the second. This palindrome preferentially assumed a 'S'-like average shape on the crystal surface of mica. The 'tail-tail' dimer instead displayed a negative curvature for the first half and a positive curvature for the second, that corresponds to a preferential average shape that is mirror image of the S-like one: we labeled it 'S*'.

Similarly to the above-mentioned case of pBR322 dimers, this result confirmed that mica could direct the deposition of DNA molecules in a face-specific fashion. Also from these dimers, inverting the directions of the half sequences produces shapes that can be obtained by only rotating the halves and so letting always the same faces adsorb preferentially on the mica surface. From the comparison of experimental data with the theoretical curvature predictions, it turns out that the mica surface preferentially interacts with the T-rich molecular face.

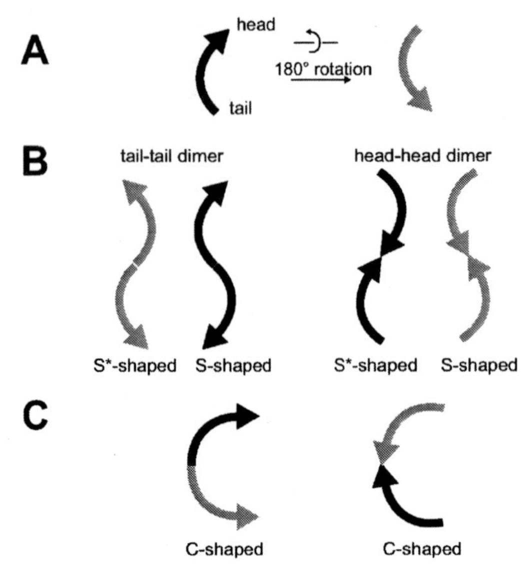

Figure 6. A) Pictorial representation of a curved DNA segment. Curvature is due to the phasing of AA-tracts along the helix. Adenine steps (and consequently Thymine steps) are preferentially segregated on one face of the molecular plane. In this scheme, the DNA helix is schematized as a curved arrow, the color of which tells what base is towards the observer: gray is for A, black for T. The orientation of the base sequence of the DNA monomer used to build the dimer can be termed as tail-head (arrowheads indicate the polarity). If the direction of the sequence is reversed, the bases toward the observer are interchanged with those complementary. B) Scheme of the tail-tail and head-head dimerization of the curved tracts, upon their deposition on mica in S-like forms: S-shaped dimers in one case and S*-shaped dimers in the other expose the same faces to the surface. The C-shaped dimers expose both the types of faces (A-rich and T-rich) to the surface. C) Scheme of the deposition of the tail-tail and head-head dimers on mica in the C-like forms.

A Few Symmetry Considerations

Only if the sequence orientation for each of the imaged molecules is known, then it would be possible to determine the face with which the molecules adsorb on the surface (on average) through the study of the average chain-curvature of the molecule ensemble. In 3D, the dyad axis which characterizes the time-averaged conformation of any palindromic DNA dimer, can be oriented along any direction of space with respect to the average plane of the curved tracts. When the dimer is flattened on a surface, only two alternative directions are allowed: parallel or perpendicular to the surface plane. In the former case, the directions of the two curvatures point towards the same side of the molecule, and a 'C'-like shape is obtained. In the latter case, the two curvatures point towards opposite directions, then an 'S'-like shape is obtained, instead (see Fig. 6).

A curve in space can be fully described by its local-curvature amplitudes and phases [42]. On the surface, the phase information is reduced to the sign of the curvature: if a segment of the curve must be rotated clock-wise to be aligned with the direction of the preceding segment, then the local curvature is defined positive. The C-like molecules will be characterized by two negative curvatures or two positive ones, depending on which end is chosen as the beginning of the molecule (and the sequence read will be the same, thanks to palindromicity). The two main curvatures in an S-like molecule will be oppositely signed instead, either with a positive followed by a negative one, a form termed 'S', or by a negative followed by a positive one, termed 'S*'. The asterisk in the terms indicates a mirror shape: because of the dyad symmetry, the two face-shapes are prochiral and expose complementary sequences. These two possibilities are the result of the adsorption of the three-dimensional dimeric molecules on either one of the two opposite faces (Fig. 6). In the case of C-shaped molecules (Fig. 6), the two faces are equivalent, instead, because within either face one half exposes a sequence complementary to that of the other half. This also implies that the two faces of the C-shaped molecules are indistinguishable and belong to the same symmetry specie. On this basis, in a curved palindromic molecule, the face (either A-rich or T-rich) that is exposed to the surface can be identified on a morphological basis from its SFM images.

On the average, a palindromic molecule of proper sequence can therefore expose to mica either an A-rich face or a T-rich one and we can identify the face preferentially exposed to the surface (if there is any) from the average molecular shape. With this kind of construct, questions about the capability of mica to recognize DNA sequences can be thus addressed. In fact, if mica interacts in the same way with different DNA sequences a dimeric molecule will deposit on the surface either with the S- or with the S*- like shape with the same probability. On the contrary, any disparity between these two sub-populations indicates that mica interacts in a different way with different DNA sequences.

We can expect that the two faces of the C-shaped molecules (that are physically equivalent) must be present on the surface in the same number and therefore their contributions to the average curvature must cancel out. The two faces of the S-shaped molecules instead are different. They will be present on the surface in the same number only if they are equally adsorbed on it. Their contributions to curvature will cancel out only if the surface interacts in the same way with both of them. A non zero curvature profile thus directly monitors the unbalance of adsorption of the sub-populations of the S and S* profiles. As a consequence, the two sigmoidal plots in Fig. 5 indicate that both dimers preferentially adsorb with one of the S-like forms respect to the other, and the faces upon which they were preferentially deposited can be recognized from the signs of the curvatures. The minimum-energy theoretical models of the two dimers [43] show that these two different shapes correspond to a deposition of both dimers where the same T-rich face of the curved tracts has been exposed to the mica surface. We could therefore infer that, because of the inversion of the sequence in their two halves (see Fig. 6) the two dimers were forced to assume two shapes that were mirror images to each other in order to achieve that preferential interaction.

We could rule out that this result could be due to an incomplete ensemble averaging. The symmetrical shape of the position-dependent curvature profiles and the equality in the amplitude of the curvature of the two halves of the molecules (that reflects the binary symmetry of the sequence) demonstrated that the sampled ensemble is statistically relevant for this system. After this internal check, the curvature profiles have been symmetrized to further improve the statistical averaging by doubling the information on the curved segment. Two additional proofs of complete statistical ensembling were achieved. First, for both the examined populations of DNA molecules the experimental local curvature modulus along the chain was found to be proportional to the corresponding standard deviation of the curvature modulus, as predicted theoretically (A. Scipioni et al., manuscript submitted for publication). Second, after having classified the shapes of all the imaged molecules according to the curvature of their halves into the three sub-populations of the S, S* and C classes of shapes, the profiles were averaged over these sub-populations, and a crossed equivalence of the half faces of the C and S forms was displayed. This indicates that the sub-populations are complete ensembles for the curvature analysis.

The same classification revealed that the preference of the mica surface for the T-rich face of the *Crithidia fasciculata* DNA is not a weak effect. In fact, for the tail-tail palindrome the molecules that assumed an 'S*' conformation were numerically 9 times as numerous as those in the 'S' class. Conversely, for the head-head dimer, the 'S' class is significantly more populated than the 'S*' class (5 times as much).

The recognition effect here reported might account for the two opposite preferential azimuthal orientations of DNA molecules on inorganic crystal surfaces that were assumed by Rhodes and Klug [44] in order to explain the periodicity of about 10.5 base pairs of the nucleosomal DNA digestion by DNase I.

This nanoscale recognition process might prove important for its nanotechnological implications. For example, we can foresee applications based on the capability of the two complementary sequences on the two faces of a curved DNA to be recognized by different nanocrystals that would be thus selected from a solution, directed to assemble and immobilized by these specific interactions.

ACKNOWLEDGMENTS. This work was supported by Programmi Biotecnologie Legge 95/95 (MURST 5%) and MURST Programmi di ricerca di Interesse Nazionale (Biologia Strutturale 1997-1999 and 1999-2001).

REFERENCES

1. Demers, L. M., Ginger, D. S., Park, S. J., Li, Z., Chung, S. W. and Mirkin, C. A. *Science*, **296,** 1836-8. (2002).
2. Montemagno, C. and Banchard, G. *Nanotechnology*, **10,** 225-231 (1999).
3. Niemeyer, C. M. *Chemistry*, **7,** 3188-3195 (2001).
4. Samorì, B., Muzzalupo, I. and Zuccheri, G. *Scanning Microscopy*, **10,** 953-962 (1996).
5. Seeman, N. C. *Angew. Chemie Int. Ed.*, **37,** 3230-3238 (1998).
6. Lindsay, S. M., Tao, N. J., DeRose, J. A., Oden, P. I., Lyubchenko, Y. L., Harrington, R. E. and Shlyakhtenko, L. *Biophys. J.*, **61,** 1570-84 (1992).
7. Samorì, B., Nigro, C., Armentano, V., Cimieri, S., Zuccheri, G. and Quagliariello, C. *Angew. Chem. Int. Ed.*, **32,** 1461-1463 (1993).
8. Henderson, E. *Nucleic Acids Res.*, **20,** 445-447 (1992).
9. Thundat, T., Allison, D. P., Warmack, R. J. and Ferrell, T. L. *Ultramicroscopy*, **42,** 1101-6 (1992).
10. Thundat, T., Warmack, R. J., Allison, D. P., Bottomley, L. A., Lourenco, A. J. and Ferrell, T. L. *Journal of Vacuum Science & Technology A (Vacuum, Surfaces, and Films)*, **10,** 630-5 (1992).
11. Murray, M. N., Hansma, H. G., Bezanilla, M., Sano, T., Ogletree, D. F., Kolbe, W., Smith, C. L., Cantor, C. R., Spengler, S., Hansma, P. K. and et al. *Proc. Natl. Acad. Sci. U S A*, **90,** 3811-4 (1993).
12. Lyubchenko, Y. L., Gall, A. A., Shlyakhtenko, L. S., Harrington, R. E., Jacobs, B. L., Oden, P. I. and Lindsay, S. M. *J. Biomol. Struct. Dyn.*, **10,** 589-606 (1992).
13. Lyubchenko, Y. L., Jacobs, B. L. and Lindsay, S. M. *Nucleic Acids Res.*, **20,** 3983-3986 (1992).

14. Karrasch, S., Dolder, M., Schabert, F., Ramsden, J. and Engel, A. *Biophys. J.*, **65,** 2437-46 (1993).
15. Feng, X. Z., Bash, R., Balagurumoorthy, P., Lohr, D., Harrington, R. E. and Lindsay, S. M. *Nucleic Acids Res.*, **28,** 593-6 (2000).
16. Fang, Y. and Hoh, J. H. *Nucleic Acids Res.*, **26,** 588-93 (1998).
17. Lyubchenko, Y. L. and Shlyakhtenko, L. S. *Proc. Natl. Acad. Sci. U S A*, **94,** 496-501 (1997).
18. Oussatcheva, E. A., Shlyakhtenko, L. S., Glass, R., Sinden, R. R., Lyubchenko, Y. L. and Potaman, V. N. *J. Mol. Biol.*, **292,** 75-86 (1999).
19. Rivetti, C., Guthold, M. and Bustamante, C. *J. Mol. Biol.*, **264,** 919-32 (1996).
20. Bustamante, C., Vesenka, J., Tang, C. L., Rees, W., Guthold, M. and Keller, R. *Biochemistry*, **31,** 22-6 (1992).
21. Hansma, H. G., Bezanilla, M., Zenhausern, F., Adrian, M. and Sinsheimer, R. L. *Nucleic Acids Res.*, **21,** 505-12 (1993).
22. Hansma, H. G., Sinsheimer, R. L., Li, M. Q. and Hansma, P. K. *Nucleic Acids Res.*, **20,** 3585-90 (1992).
23. Vesenka, J., Guthold, M., Tang, C. L., Keller, D., Delaine, E. and Bustamante, C. *Ultramicroscopy*, **42,** 1243-9 (1992).
24. Hansma, H. G., Vesenka, J., Siegerist, C., Kelderman, G., Morrett, H., Sinsheimer, R. O., Elings, V., Bustamante, C. and Hansma, P. K. *Science*, **256,** 1180-1184 (1992).
25. Weisenhorn, A. L., Gaub, H. E., Hansma, H. G., Sinsheimer, R. L., Kelderman, G. L. and Hansma, P. K. *Scanning Microscopy*, **4,** 511-516 (1990).
26. Rabke, C. E., Wenzler, L. A. and Beebe, T. P., Jr. *Scanning Microscopy*, **8,** 471-80 (1994).
27. Bezanilla, M., Manne, S., Laney, D. E., Lyubchenko, Y. L. and Hansma, H. G. *Langmuir*, **11,** 655-659 (1995).
28. Hansma, H. G. and Laney, D. E. *Biophys. J.*, **70,** 1933-9 (1996).
29. Han, W., Dlakic, M., Zhu, Y. J., Lindsay, S. M. and Harrington, R. E. *Proc. Natl. Acad. Sci. U S A*, **94,** 10565-70 (1997).
30. Han, W., Lindsay, S. M., Dlakic, M. and Harrington, R. E. *Nature*, **386,** 563 (1997).
31. Xu, Y. C. and Bremer, H. *Nucleic Acids Res.*, **25,** 4067-71 (1997).
32. Pietrasanta, L. I., Schaper, A. and Jovin, T. M. *Nucleic Acids Res.*, **22,** 3288-92 (1994).
33. Schaper, A., I., P. L. and Jovin, T. M. *Nucleic Acids Res.*, **21,** 6004-6009 (1993).
34. Fang, Y. and Hoh, J. H. *FEBS Lett.*, **459,** 173-6 (1999).
35. Shlyakhtenko, L. S., Potaman, V. N., sinden, R. R., Gall, A. A. and Lyubchenko, Y. L. *Nucleic Acids Res.*, **28,** 3472- (2000).
36. Zuccheri, G., Dame, R. Th., Aquila, M., Muzzalupo, I. and Samorì, B. *Appl. Phys. A*, **66,** S585-9 (1998).
37. Theveny, B. and Revet, B. *Nucleic Acids Res.*, **15,** 947-58 (1987).
38. Zuccheri, G., Scipioni, A., Cavaliere, V., Gargiulo, G., De Santis, P. and Samorì, B. *Proc. Natl. Acad. Sci. USA*, **98,** 3074-3079 (2001).
39. Marini, J. C., Levene, S. D., Crothers, D. M. and Englund, P. T. *Proc. Natl. Acad. Sci. USA*, **79,** 7664-7668 (1982).
40. Griffith, J., Bleyman, M., Rauch, C. A., Kitchin, P. A. and Englund, P. T. *Cell*, **46,** 717-24 (1986).
41. De Santis, P., Palleschi, A., Morosetti, S. and Savino, M., Structure and dynamics of nucleic acids. Pergamon Press, Elmsford, NY, (1986), pp. 31-49.
42. De Santis, P., Palleschi, A., Savino, M. and Scipioni, A. *Biochemistry*, **29,** 9269-73 (1990).
43. Crothers, D. M. *Proc Natl Acad Sci U S A*, **95,** 15163-5 (1998).
44. Rhodes, D. and Klug, A. *Nature*, **286,** 573-8. (1980).

MANIPULATION BY AN ELECTRIC FIELD

Molecular Manipulation of DNA and Its Applications

Masao Washizu

Department of Mechanical Engineering, The University of Tokyo
7-3-1 Hongo, Bunkyo-ku, Tokyo, 113-8656, Japan
washizu@washizu.t.u-tokyo.ac.jp

Abstract. A method for molecular manipulation of DNA has been developed. The method uses microfabricated electrode system to create high-intensity high-frequency field in DNA solution, by which each DNA molecule is stretched to a straight shape, aligned with one end immobilized onto the electrode edge. Once stretched and immobilized, one has an access to, and can apply operations to any desired location on the strand. It has been shown that aimed portion of the stretched DNA can be dissected and picked up. Immobilization in some case cause steric hindrance and hamper the interaction of enzymes with the DNA. To prevent such hindrance, a microdevice has been developed which enables the anchoring of stretched DNA with its molecular termini, leaving middle part free. Using the device, enzymatic molecular surgery of DNA is demonstrated by pressing a DNA-cutting-enzyme coated microparticle against the stretched DNA. Real-time observation of DNA enzymes moving along DNA strand is also made using the device.

INTRODUCTION

There are two ways towards DNA-based molecular construction, bottom-up and top-down. The bottom-up approach, drawing more and more attention these days, makes use of the self-assembling of the molecules, and features massive parallelism. On the other hand, top-down approach is the extension of micro fabrication/machining, and expected to play a role in interfacing molecular nano-world and our macro-world, for instance the circuit connections of molecular electronics devices. One key technology in top-down DNA molecular construction may be the method to position and align DNA molecules, however, up to now, few research effort has been invested for such physical manipulation of DNA molecules. The author and his co-workers has been working with bio-manipulation for years, and found that electrostatic field effects in microsystems can be a powerful tool for such purposes. This paper describes the principle together with some experimental results.

AC FIELD EFFECTS FOR DNA MANIPULATION – DIELECTROPHORESIS AND ELECTROSTATIC ORIENTATION

DNA has the double-stranded helical structure with the diameter of 2 nm and the length of 0.34 nm per base pair (1 μm for 3 kbp). Such a long string-like molecule in

CP640, *DNA-Based Molecular Construction: International Workshop,* edited by W. Fritzsche
© 2002 American Institute of Physics 0-7354-0095-4/02/$19.00

water solution takes randomly coiled conformation due to thermal agitation. In order to stretch DNA to straight conformation and align onto a solid surface, we use ac field effects, namely, dielectrophoresis [1] and electrostatic orientation. Fig.1 depicts these effects. Fig.1-a) shows an electrically neutral particle placed in a non-uniform electrostatic field produced by a pin-plate electrode system. The particle polarizes, and an equal amount of positive and negative charge is induced on the surface of the particle. Coulombic force F = q E works on these charges, and if the field is uniform, these two forces cancel each other to yield no net force. But when the field is non-uniform, the two forces are different in magnitude and/or direction, and net force results. Fig.1-a) depicts the case where the polarizability of the particle is larger than that of the medium, when a positive charge appears on the downstream side of the field line, and a negative charge on the other side. Because the field is stronger at the left side of the particle in the figure than the right side, the particle is moved towards the pin electrode where the field is stronger. The motion towards stronger field is termed positive dielectrophoresis (positive DEP). If the polarizability of the particle is smaller than that of the medium, the induced charges will be of opposite polarity to what is depicted in the figure, and the particle is moved toward where the field is weaker (negative DEP). Positive DEP is more often observed in biological DEP than the negative.

Fig.1-b) depicts an elongated object in an electrostatic field. The Coulombic force on the polarization charge rotates the object, until its longest axis becomes parallel to the field. The phenomenon is termed electrostatic orientation. (It is known, in a narrow frequency range, the longest axis can be orientated perpendicular to the field [2], but this case is rarely observed experimentally.) When the object is a flexible fiber like DNA, the electrostatic orientation occurs in every part of the object, and as a result, it is stretched to a straight shape, parallel to the field.

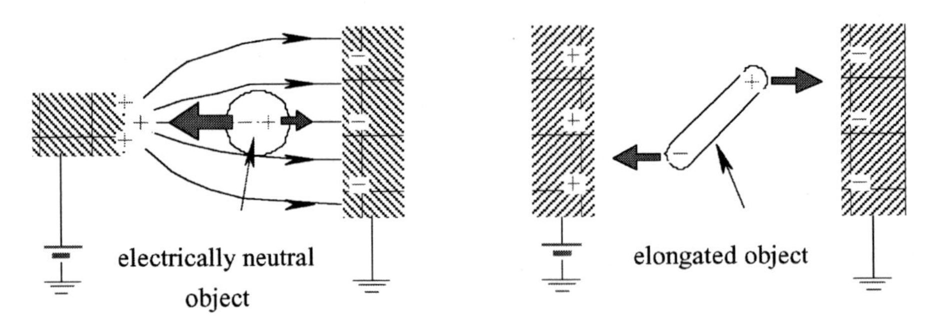

FIGURE 1. AC field effects for DNA manipulation – dielectrophoresis and electrostatic orientation

These electrokinetic effects are the result of the interaction between the externally applied field and the induced charge. In an ac field, the direction of the field alternates in the next half cycle, but so far as the polarization can follow the field alternation, the induced charge also changes its sign, and the direction of force remains unchanged. Therefore, the effects are effective in ac field as well as in dc field. This definitely is advantageous for the use in water solution. By using high frequency ac field, electrochemical reactions such as electrolysis at the electrode/water interface can be avoided, so that electrodes can be embedded in water phase. The electrode gap can be made arbitrarily small, so that very high intensity field can be obtained with moderate power supply voltage. By confining the high-field region to a small volume, having large surface-to-volume ratio, heat can effectively be removed, so a high-intensity field can be created without excessive temperature rise.

Fig.2 shows the alignment method of DNA using the electrokinetic effects. A pair of thin-film metal electrode is patterned onto a substrate, onto which DNA solution is fed, and then covered with a cover slip. The electrode is energized to 1 MV/m (100 V across 100 micron) by a high frequency (1MHz) power supply. The electrostatic orientation occurs, and the DNA strand is stretched to a straight shape. Then dielectrophoresis pulls the strand towards the electrode edge where the field is most intense, until one molecular end touches the electrode. The touching end is permanently anchored when electrochemically active metal such as aluminum is used as the electrode material. The stretching is instantaneous, and the whole process completes within several seconds. We named the process "electrostatic stretch-and-positioning of DNA"[3-4]. The method enables the stretching of individual DNA strands simultaneously, and aligning them with one end in line. The density obtainable with this method is 6 – 7 DNA strands per 1 μm of electrode contour.

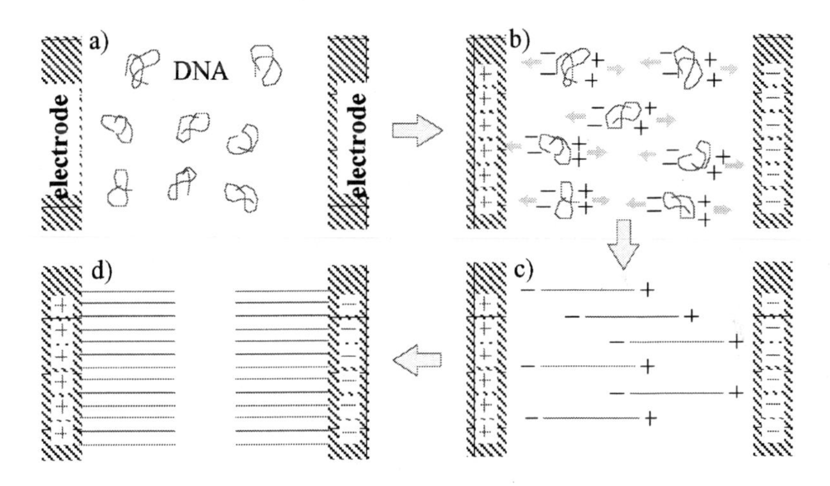

FIGURE 2. Electrostatic stretch-and-positioning of DNA

The mechanism of DNA anchoring onto the electrode edge is still an open question. The facts that a) the anchoring is so strong that DNA strand would rather break than coming off the electrode, b) the anchoring is stronger when electrochemically active metal such as aluminum is used as the electrode material than gold or platinum, c) the anchoring is stronger with fresh vacuum-evaporated Al, and becomes weaker as it gets old and surface-oxidized, d) the anchoring is restored if old Al electrode is treated by alkaline solution such as KOH, may suggest covalent bond is electrochemically formed between DNA and the electrode.

The mechanism of DNA polarization is the counter-ion polarization [5]. DNA has negatively charged phosphate backbones, which attracts positive ions in solution to form counter-ion cloud Debye layer around the strand (fig.3). When the field is applied, the counter-ion moves along the strand, and excess positive ion is accumulated at the downstream side of the field, while negative backbone charge is exposed at the upstream side, hence the polarization occurs.

In order to avoid Joule heating under the high intensity field required to stretch out DNA (1MV/m), the conductivity of the medium must be low. In addition, it is the polarizability of the counter-ion cloud relative to that of the medium that determines the orientation effect. To keep the polarizability of the medium low, the medium conductivity again must be kept low. Therefore, the electrostatic stretch-and-positioning of DNA in most case is performed in deionized water, or at least in the medium of <1mM. When thicker buffer solution is required, for instance in the case where enzymatic reaction is involved, DNA is stretched and positioned first, and then the medium is replaced.

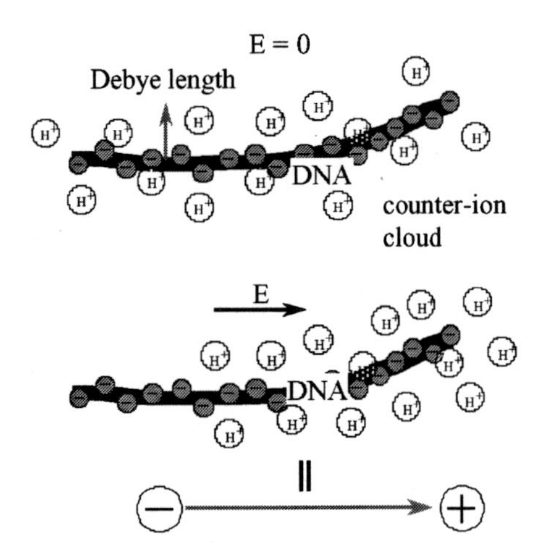

FIGURE 3. Mechanism of DNA polarization

MICRO-DISSECTION OF STRETCHED DNA

The stretch-and-positioning technique can be used for dissection and acquisition of aimed position of a DNA strand. The dissection of the stretched DNA itself can be done mechanically using a sharp stylus as a cutting knife, but picking up the dissected fragment is not very easy. For this purpose, we have developed a method and a microdevice based on sacrificial etching, which is schematically depicted in fig.4 [6]. The device consists of a glass substrate, onto which a sacrificial layer, DNA carrier layer, and a pair of electrodes are deposited. DNA is electrostatically stretched and immobilized onto the carrier layer with one of its molecular end aligned on the electrode edge. Using an AFM stylus as a knife, an aimed portion of the DNA together with the carrier layer is dissected. By dissolving the sacrificial layer, the DNA fragment on the piece of carrier is recovered onto a membrane filter. The carrier piece is then melted to obtain DNA fragments in solution.

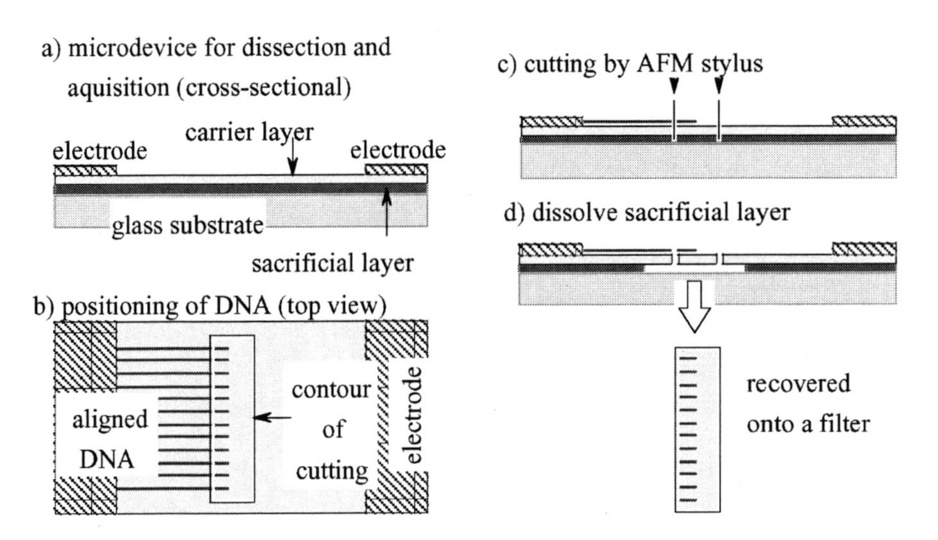

FIGURE 4. Dissection and acquisition of aimed portion of DNA using a microdevice

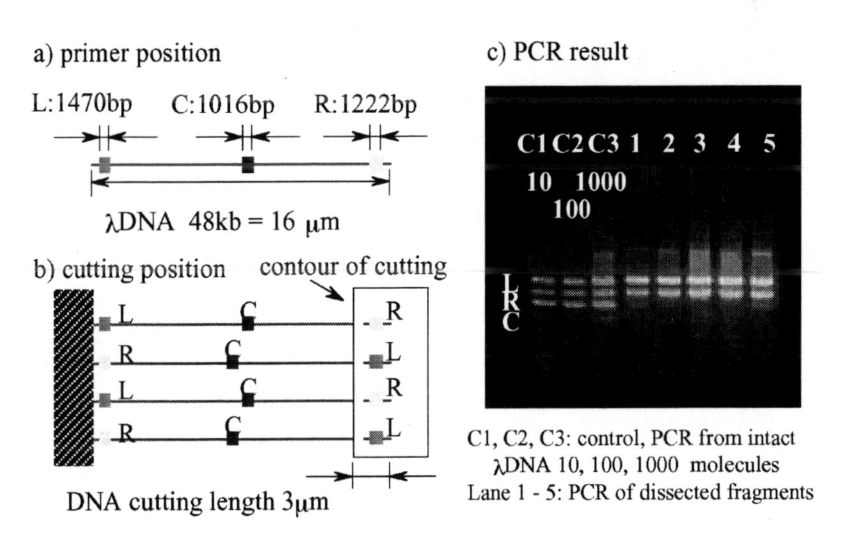

FIGURE 5. Experimental demonstration of the dissection and acquisition

An experimental demonstration of the physical DNA dissection is performed using λDNA as the sample. In order to prove that the desired position is dissected and picked-up, three primers are prepared, as depicted in fig.5 a). They correspond respectively to the sequences near the left end of λDNA (denoted L), near the right end (R), and approximately at the center (C), all about 1kb in length. Because DNA is electrically symmetrical, the electrostatic positioning yields a mixture of one orientation and the other, as shown in fig.5 b). If aligned λDNA is cut, say at 3 μm from the end, and successfully picked up, the PCR product should contain the sequence L and R, but not C (fig. b). On the other hand, if the sequence C is detected by PCR, it is an indication that unwanted DNA fragments are coming in.

Fig.5 c) shows the electrophoresis of the PCR result. Lanes C1-C3 are positive control, starting from 10, 100 and 1000 non-cut λDNA molecules respectively. In these lanes, the three bands, from top to bottom L, R, and C, are seen. Lanes #1 through #5 are PCR of the carrier piece. All show the bands from L and R, and not C.

These results demonstrate the successful mechanical dissection and recovery of DNA. From the brightness of the bands in comparison with that of C3, the number of dissected molecules is estimated to be several thousand per run, which is adequate for the use in biochemical processes, such as sequencing.

DNA-PROTEIN INTERACTION

The active site of an enzyme in most case is inside a molecular cleavage. Adsorption of DNA onto a solid surface, as in the case of fig.4, often hampers DNA from entering deeply enough into the active site, and enzyme is unable to work. In order to obtain stretched DNA that does not cause such a steric hindrance, a micro electrode system is developed, which is depicted in fig.6.

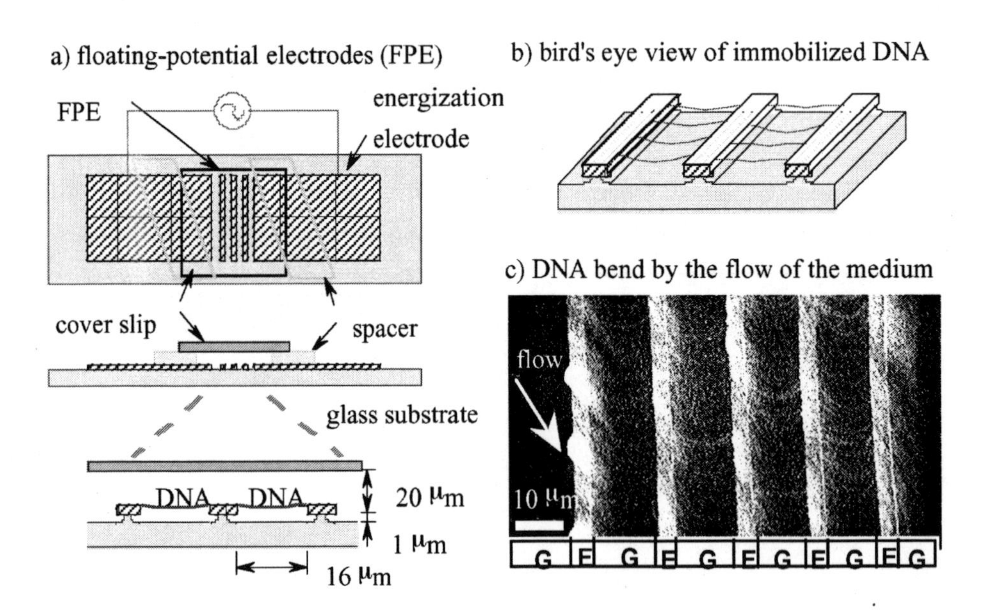

FIGURE 6. Floating-potential electrode system for immobilization of DNA at molecular ends.

The system consists of a pair of energizing electrodes on a glass substrate, and a few thin strips of aluminum having no electrical connection, which we call floating-potential electrodes (FPE). The spacing between the floating-potential electrodes are made slightly smaller than the length of DNA to be immobilized, and the glass surface between the electrodes are etched down about 1μm. When the outer electrodes are energized, DNA is stretched and pulled towards the edge of the electrodes. When one molecular end of DNA touches an electrode, it is anchored. At this moment, the other end extends to the edge of adjacent electrode, and is dielectrophoretically pulled-in to be anchored. Because the glass surface is lower, DNA is held free except at both ends. The function of FPE here is to deform electrostatic field to create field maxima for DEP trapping of DNA. These electrodes are better left to the floating potential: when electrodes are connected to power supply with low impedance, charge injection creates jet-like flow at the very vicinity of the electrode edge, which hampers the approaching of DNA ends.

Fig.6c) is the photograph of DNA bridging over the FPE's. Here, a flow of the medium as shown by the arrow is intentionally created, by which DNA strands are bent downward. This proves that the middle part of DNA is not adsorbed on the solid surface, and that the anchoring is strong enough to withstand the hydrodynamic drag.

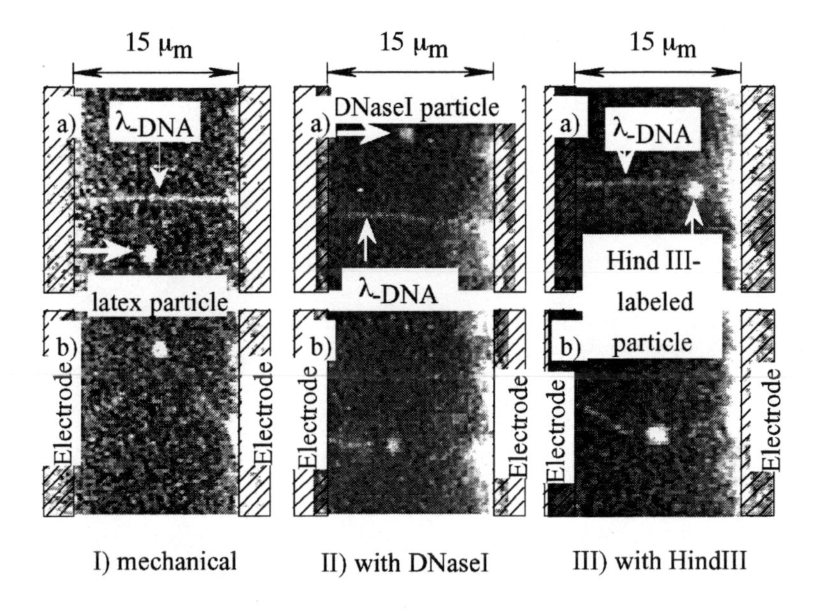

FIGURE 7. Molecular surgery with enzyme-immobilized microparticle
I) Mechanical Breakage of DNA
a) Immobilized DNA and a particle having no enzymes.
b) Breakage of the DNA occurs when DNA is mechanically elongated to 150%.
II) Molecular Surgery with DNaseI-labeled particle
a) Before contact
b) DNA is cut as soon as the particle touches.
III) Molecular Surgery with HindIII-labeled particle
a) The particle is already in touch with the DNA, but DNA is not cut.
b) The particle is moved keeping contact with DNA, and at some point, DNA is cut.

By using the electrode system, molecular surgery of DNA with enzyme-immobilized particle is demonstrated [7], whose time snap is shown in the photo of fig.7. DNA cutting enzyme is immobilized on a microparticle having the diameter of

1μm. It is grasped with a laser tweezers under a microscope, and is pressed against the immobilized DNA. By doing so, one can specify the location of the enzymatic reaction, namely at the contact point. This is in contrast to the conventional enzymatic reaction in solution, in which the location of the reaction is only statistically predictable.

Fig.7-I is the result with a plain particle i.e. the particle having no enzyme, which is used as a reference to determine if the DNA cutting is mechanical or enzymatic. In this case, DNA has to be elongated to 1.5 times of its original length before the mechanical breakage occurs. In fig. 7-II DNaseI, the enzyme which cuts DNA regardless of the base sequence, is immobilized on the particle. In this case, the cutting occurred as soon as the particle touches the DNA. Fig. 7-III is the case for a particle having a restriction enzyme on the surface. HindIII used in this experiment recognizes a sequence of 6-bases, and has 7 restriction sites on λDNA used in this experiment. When this enzyme is immobilized on the particle, contact with DNA does not always result in cutting. It is as it should be, because the particle is most likely to hit locations other than restriction sites. Therefore, the particle is moved along DNA keeping the contact. The particle apparently overlooks some of the restriction sites, but if the scanning is repeated for a few times, DNA is cut as shown in fig.7-III-b). The cutting position in agreement with the restriction sites is observed.

VISUALIZATION OF ENZYME DYNAMICS ON DNA AND ITS APPLICATION TO OPTICAL MAPPING

The immobilization of DNA using floating potential electrodes provides an ideal test-bed for the observation of DNA-protein interaction [8], which can be used for optical mapping of specific sequences and investigation of the dynamic process of the interaction.

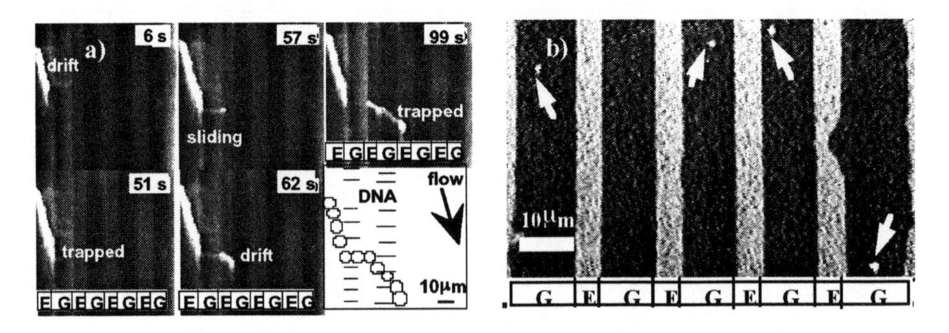

FIGURE 8. Sliding motion of EcoRI on λDNA (a) and final trapping positions (b)

Fig.8 a) shows the sliding motion of a fluorescence-labeled restriction enzyme EcoRI along a DNA strand. The enzyme first non-specifically binds to any part, and then slides along the strand like a monorail. Through one-dimensional relocation by the sliding, the enzyme can more efficiently reach the restriction site compared with simply relying on 3-dimensional diffusion.

In fig.8 b) is shown the location of the enzyme finally trapped on DNA strands (indicated by arrows). Here, the observation is made in the absence of Mg^{++}, the cofactor necessary for cleavage of DNA, and optional for binding. The trapped locations are in agreement with the restriction enzyme, showing the potential of the method for optical mapping of DNA.

CONCLUSIONS

Physical molecular manipulation method of DNA based on microfabrication is developed, and as examples of its application, 1) dissection of aimed portion of DNA, picking up and recovery, 2) molecular surgery with an enzyme-immobilized probe, 3) real-time observation of the binding of restriction enzymes to stretched DNA, are experimentally demonstrated. The spatial resolution provided by the electrostatic stretch-and-positioning method is expected to open a way for bio-nanotechnology.

ACKNOWLEDGMENTS

The author would like to thank Dr. Hiroyuki Kabata of Tokyo Univ., Mr. Kurosawa of Advance Co., Dr. Takatoki Yamamoto of Tokyo Univ., Prof. Nobuo Shimamoto of the Institute of Genetics for collaborations. This work is in part supported by BRAIN (Seiken-Kiko) Research and Development Program for New Bio-industry Initiatives, NEDO (Proposal-Based R&D Program 97S07-005), the Ministry of Education (Kakenhi 11450103, 12555076, 12030216), Micromachine Center, and Advance Co.

REFERENCES

1. Pohl, H. A., Dielectrophoresis, Cambridge University Press (1978)
2. Jones, T.B., Electromechanics of Particles, Cambridge University Press (1995)
3. Washizu, M. and Kurosawa, O: "Electrostatic Manipulation of DNA in Microfabricated Structures", IEEE Trans. IA, Vol.26, No.6, p.1165-1172 (1990)
4. Washizu, M., Kurosawa, O, Arai, I., Suzuki, S. and Shimamoto, N.: "Applications of Electrostatic Stretch-and-positioning of DNA"_ IEEE Transaction IA. Vol.31_ No.3_ p.447-456 (1995)
5. Suzuki, S., Yamanashi, T, Tazawa, S., Kurosawa, O and Washizu, M: "Quantitative analysis on electrostatic orientation of DNA in stationary AC electric field using fluorescence anisotropy", IEEE Transaction IA, Vol.34, No.1, p.75-83 (1998)
6. Kurosawa, O., Okabe, K. and Washizu, M.: "DNA analysis based on physical manipulation", Proceedings of the thirteenth annual international conference on Micro Electro Mechanical Systems (MEMS2000), p.311-316 (2000)
7. Yamamoto, T., Kurosawa, O., Kabata, H., Shimamoto, N. and Washizu, M.: "Molecular surgery of DNA based on electrostatic micromanipulation", IEEE Transaction IA, Vol.36, No.4, p.1010-1017 (2000)
8. Kabata, H., Okada, W. and Washizu, M.: "Single-Molecule Dynamics of the Eco RI Enzyme using Stretched DNA: Its Application to In Situ Sliding Assay and Optical DNA Mapping", Jpn. J. Appl. Phys. Vol.39, p.7164-7171 (2000)

Oriented Immobilization of Single DNA Molecules as a Nanostructuring Tool

Frank F. Bier, Nenad Gajovic-Eichelmann, Ralph Hölzel

Fraunhofer Institute for Biomedical Engineering, Dept. of Molecular Bioanalytics & Bioelectronics, Arthur-Scheunert-Allee 114-116, D-14558 Bergholz-Rehbrücke, Germany

Abstract. To facilitate the directed positioning of molecules on a surface the use of nucleic acids is proposed. For nanostructuring with DNA the deposition of each molecule has to be directed. Two steps are necessary: First, the DNA has to be stretched and second the lengthy molecule has to be fixed at each end in a specific manner. For stretching DNA of several micrometer length AC fields are applied, for anchoring the ends in a specific manner the oligo-tag method is applied using different immobilization techniques.

INTRODUCTION

Nucleic acids have a well characterized structure that might be exploited for physical partitioning of a certain space or a surface. Since single molecule manipulation has gained much interest over the last years, methods are now at hand to construct and observe surfaces, that have been ordered on the molecular scale. Concurrently, self-assembly has been used for the building of nano-constructs [1]. For the construction of long range physical structures with nanometer scale addressability molecular manipulation has to be combined with site directed immobilization of anchoring entities. The basic notion of the present work is to anchor a long DNA molecule by specific binding at least at each end and thus to bridge a gap of several micrometer. Using the specific sequence of the DNA each site in the gap may be addressed in a raster at a resolution of 0.34 nm, which is the distance between bases along the backbone of the DNA if it is in its usual helical B-form. Two steps are necessary to facilitate such constructs: First the DNA has to be stretched, second the ends of the DNA have to be fixed on the surface in a specific manner.

To achieve the first goal AC fields have been employed succesfully following a method introduced by Washizu [2]. Site directed immobilization of the ends of a long DNA strand has been achieved by use of the oligo-tag method [3]: Oligomers that are complementary to the ends of the long DNA molecule are immobilized at distinct points. For stability reasons doublestranded DNA is preferred as the bridging molecule. Therefore a DNA-analogue, the peptide nucleic acid (PNA) introduced by Nielsen and coworkers [4], was employed, since PNA-oligomers are able to invade into the doublestrand and to bind in a base specific manner.

The basic notion of our concept is sketched in Figure 1.

CP640, *DNA-Based Molecular Construction: International Workshop,* edited by W. Fritzsche
© 2002 American Institute of Physics 0-7354-0095-4/02/$19.00

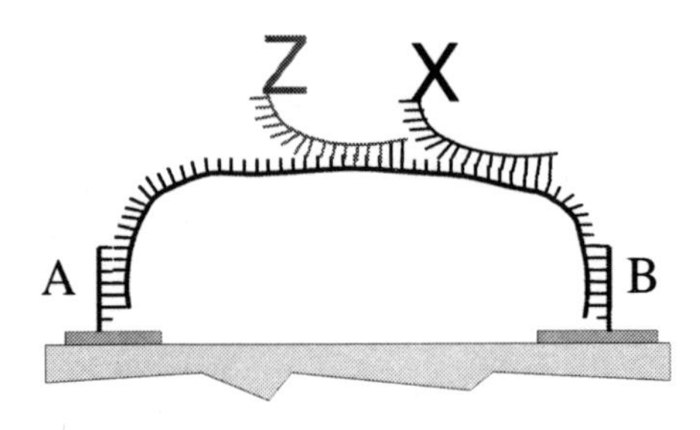

FIGURE 1. Proposed method to direct and fix molecules with DNA with defined orientation on a surface. Scheme is not to scale. At distinct areas different oligo-nucleotides, A and B, are immobilised. A long single-stranded DNA polynucleotide binds by hybridisation with its two ends complementary to A and B, respectively. Molecules X and Z are directed to well defined positions by their oligo-tag.

DIELECTROPHORETIC CONCENTRATION AND ALIGNMENT OF DNA MOLECULES

The persistence length of double stranded DNA [5] is much shorter than the physical length of the linearized M13 molecule being 2.4 µm. This means that the molecule usually is coiled. To get it into a stretched configuration an oriented physical force is needed. A number of authors apply a laminar flow of the surrounding fluid [6], which, on the first sight, is a simple procedure. However, proper control and reproducibility of experimental conditions is not an easy task. Similar problems arise when using magnetic microbeads coupled to the DNA and applying magnetic fields [7]. As an alternative, electric fields can produce the necessary lateral forces[8, 9]. For this purpose planar interdigitated electrodes (IDE) on a glass or quartz substrate can be used (Fig. 2). The electrode structures are produced by standard photolithography as in semiconductor production. The electrodes are powered by AC sinusoidal signals, typically around 1 MHz from a laboratory function generator. Due to the minute electrode thickness the resulting three dimensional electric field is strongly inhomogeneous [10]. This leads to an attracting force towards the region of highest field strength, which is close to the electrode edges. This effect is called dielectrophoresis and has been exploited for decades e.g. for the separation of biological cells [11]. For a long time it had been assumed that this effect is only of importance to objects of at least microscopical size because of disturbing Brownian motion. However, Washizu et al. [2] have been the first to demonstrate that with the help of electrodes with gap widths of some 10 µm it is possible to attract also macro molecules like DNA. Whilst it is quite helpful to concentrate solutes from the bulk solution at well defined positions just by electrical means, for this work the main advantage of AC electric fields is the stretching force acting on the particle. This force can easily be controlled by adjusting frequency and amplitude of the external voltage source. Above that molecules are automatically oriented parallel to the field lines. A drawback of dielectrophoresis is the need for solutions of low electrical conductivity in order to avoid turbulances by thermal heating as well as electrohydrodynamic effects [12]. Fortunately, DNA peristence length is maximal at low ionic strengths.

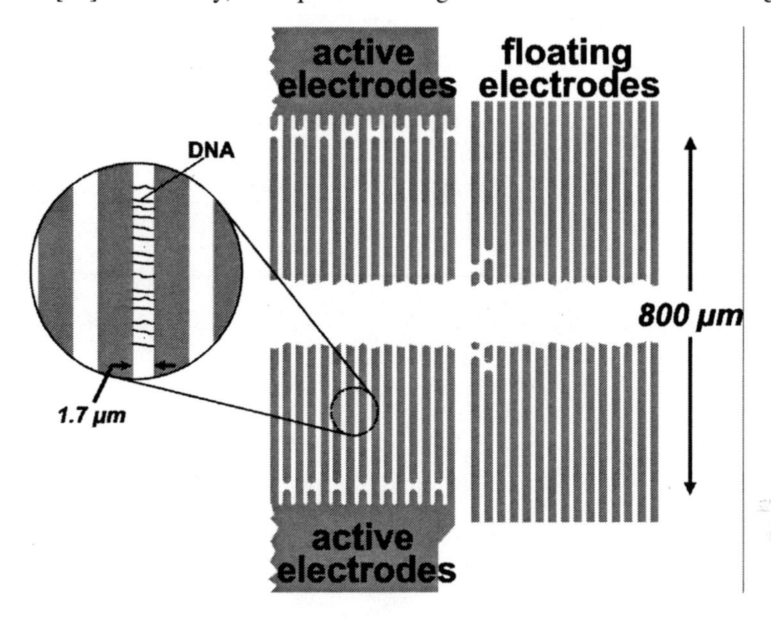

FIGURE 2. Layout of the interdigitated electrodes. Parts of three electrode sets are shown. The active IDE structure consists of 35 pairs of aluminium strips each 800 μm long and 2.3 μm wide, separated by a 1.7 μm gap. Floating electrodes serve as a control. DNA is dielectrophoretically attracted into the interelectrode gap.

To demonstrate the local concentration of molecules, linearized M13 dsDNA was dissolved in ultra pure water (200 ng/μl) and stained with the intercalating fluorescent dye PicoGreen (Molecular Probes). 1 μl of the solution was placed on interdigitated aluminium electrodes on a quartz substrate with a gap width of 1.7 μm and observed immediately with an epifluorescence microscope. Electrodes were prepared from commercially available surface acoustic wave (SAW) filters (Siemens/Matsushita R 2633). These filters consist of three IDE pairs. Only one was used for field generation, the others served as a control (Fig. 2). At 1 MHz and 2.5 V amplitude fluorescence in the gaps between the active electrode pair clearly increased over a period of about an hour (Fig. 3), whilst background fluorescence at the inactive electrodes remained close to zero. A quantitative examination of the fluorescence signal (Fig. 4) reveals an asymptotic exponential increase of DNA concentration in the interelectrode space. The pronounced fluorescence in the peripheral gap at the lower right after 12 minutes and later can be interpreted as a large supply of DNA still being available for collection in the electrodes' surrounding. Interestingly, there are some gaps with comparatively low fluorescence, and around 12 and 19 minutes fluorescence seems to be concentrated over limited stretches within some gaps. The reason for this is unclear at present. It might be a non-linear effect caused by disturbance of the field by DNA molecules already been collected at the electrodes or by aggregation of molecules. Aggregation has been observed for dielectrophoretic collection of biological cells [11] and is interpreted as a mutual dielectrophoretic attraction.

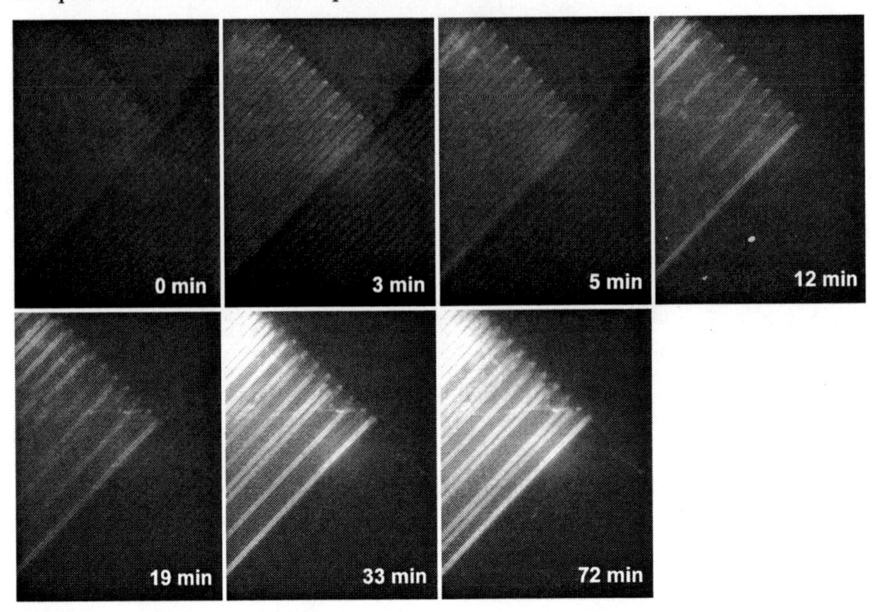

FIGURE 3. Fluorescence micrographs of dielectrophoretically collected M13 DNA. Excitation signal 2.5 V at 1 MHz. DNA was labelled with PicoGreen. The same field of view was acquired 7 times. Reprinted from [9], with permission from Elsevier Science.

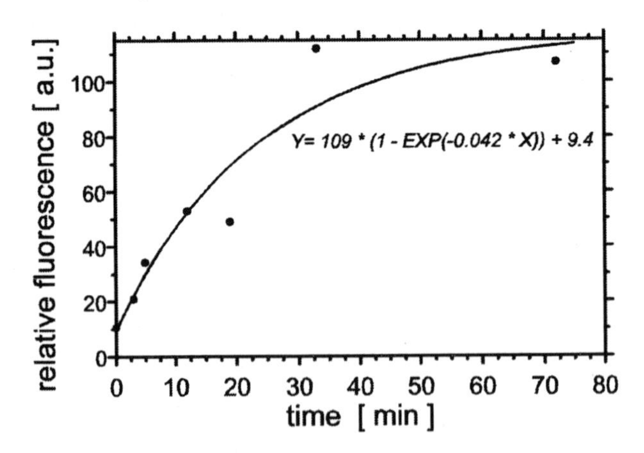

FIGURE 4. Time course of dielectrophoretic collection of M13 DNA. Fluorescence intensity was taken from Fig. 3. The background level at the lower right corner of each image was taken as a reference. Reprinted from [9], with permission from Elsevier Science.

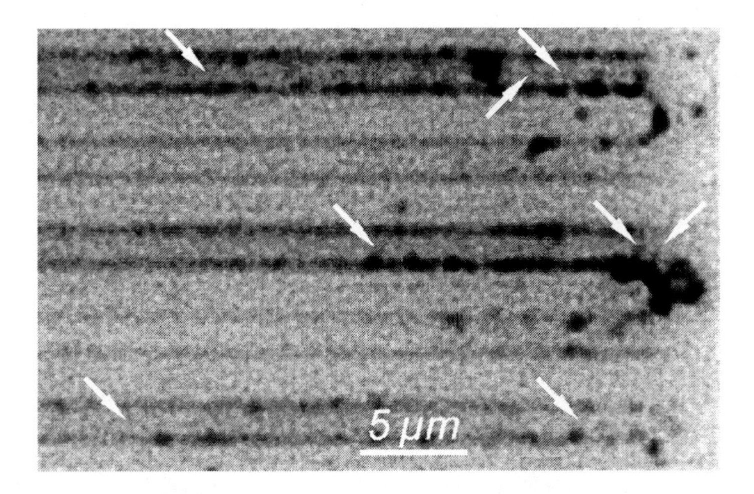

FIGURE 5. Orientation of single DNA molecules. The image has been inverted for clearer representation. M13 DNA was stained with PicoGreen. Arrows point to discernible molecules. Field frequency 1 MHz at 3 V amplitude.

In order to prove the alignment of individual DNA molecules [13] the experiment was repeated at much lower DNA concentration (Fig. 5). In this case thin straight lines are discernible bridging the interelectrode gap. These lines are quite faint, presumably because Brownian motion still smears out the image at the needed exposure times of about half a second. In this experiment molecules appear larger than the 1.7 µm gap and, hence, are not parallel to the field lines.

DETECTION OF VECTORIAL MOLECULAR ORIENTATION

The orientation of individual DNA molecules and their degree of stretching can be investigated by the rather simple means of optical fluorescence microscopy and staining with dyes that directly bind to the molecule like acridine orange, PicoGreen or YoYo [14]. Unfortunately, this method does not allow to deduce vectorial information, i.e. the knowledge about which is the 5' end and 3' end, resp. For this purpose a fluorescent sequence specific probe would be helpful. However, integration of fluorescent bases or similar compounds into the DNA molecule would render an extremely weak fluorescent signal, that would be very difficult to detect in single molecule experiments. A more promising approach is the use of microbeads of around 0.1 µm size which are loaded with a fluorescent dye and which are clearly visible in the fluorescence microscope, even under real time conditions. These microbeads can be coupled to oligonucleotides being complementary to freely selectable sequences on the DNA. Still, there are reasons to prefer double stranded DNA to single stranded DNA for nanostructuring. Thus the problem arises that oligonucleotides do not tend to hybridize to double stranded DNA. We solved this problem by using peptide nucleic acids (PNA). These are synthetic molecules with the DNA's phosphate backbone being replaced by a peptide chain. The lack of the strongly negatively charged phosphates allows such molecules not only to hybridize with single stranded DNA but also to bind to double stranded DNA [4]. At the same time sequence specificity can be retained, although obeying more complicated rules than simple DNA-DNA hybridization. PNA can bind to double stranded DNA in several ways. For example extremely stable 2:1 complexes are formed by the interaction of homopyrimidine PNA probes with homopurine sequences of double stranded DNA. These rules make it more complicated to find a PNA sequence that unambiguously binds to, e.g., a DNA site close to the 3' end.

For these investigations two types of fluorescent microbeads were used (Molecular Probes). Beads with a diameter of 40 nm and an emission maximum at 605 nm, already supplied with a streptavidin coating, were tagged with a 5'-biotinylated PNA 15mer probe. Larger, amino coated beads of 200 nm diameter with an emission around 515 nm and much brighter fluorescence were first coated with streptavidin and then tagged with a second 5'-biotinylated 15mer PNA probe.

These preparations were mixed with the M13 DNA solution and 2 µl of this mixture was pipetted onto an interdigitated electrode structure. Because of the ion strength being increased somewhat by the micro bead suspension this time a higher frequency of 2 MHz with 3 V amplitude was applied to the electrodes. The result after one hour is shown in Fig. 6. Resolution of the image has been improved afterwards by deconvolution by a Richardson-Lucy algorithm (Vision Pro, DTA, Italy). Both bead types can be differentiated by their fluorescence intensity. Although bead colours could be distinguished by direct observation through the eyepiece, the CCD camera (Kappa PS 2 C) only allowed black-and-white documentation.

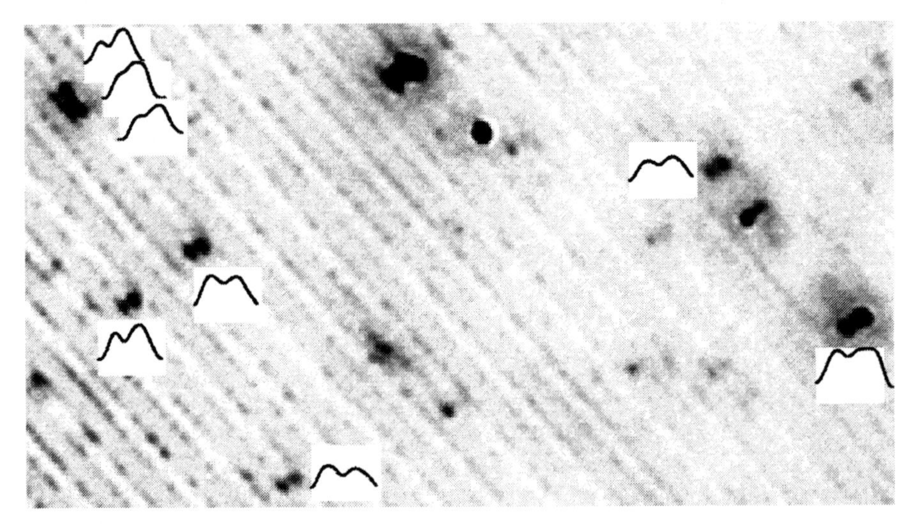

FIGURE 6. Orientation of single DNA molecules labelled with fluorescent microbeads. The image has been inverted for clearer representation. M13 DNA was specifically labelled at defined sites with two types of fluorescent microbeads using PNA. Unsymmetric pairs of beads are clearly visible. The fluorescence intensity profile along the symmetry axis of each pair is sketched close to the corresponding beads. The darker beads are close to the 3' end of the DNA.

The image clearly shows that most fluorescent objects are present in pairs. They are located in the interelectrode gap and their symmetry axis lays perpendicular to the electrode edges. From this follows that DNA molecules have been aligned bridging the electrode gap parallel to the electric field lines. Although this was expected from the results with DNA stained with an intercalator, it shows that dielectrophoretic DNA manipulation is not influenced by the staining procedure but is a property of the DNA molecule alone. Looking at the intensity profile of the bead pairs reveals that in all cases in which the pairs are positioned in the electrode gap the brighter bead is positioned exactly halfway between the electrodes. This agrees very well with its expected position on the M13 DNA, the position of which is well known. From the sequence of the PNA which had been coupled to the 200 nm bead a position results, which is distant from the 5' end 48% of M13's total length. The smaller, darker bead is expected at 79% from M13's 5' end, which again fits well to the image. These positions give a theoretical distance between both beads of 0.75 µm, whilst the intensity profiles lead to a somewhat larger average distance of 1.2 µm. From the mutual orientation of both bead types the vectorial orientation of the DNA can be deduced. Here the majority of DNA molecules is oriented with their 3' end to the lower left of the image. Whether this is a true effect or just a consequence of poor statistics remains to be studied further.

These results show, that it is possible to prove the direction of individual DNA molecules by classical optical microscopy using fluorescent microbeads that are coupled to PNA-probes.

VECTORIAL IMMOBILIZATION OF SINGLE DNA MOLECULES

To get full access to the regulative power of the DNA-construct, it should be oriented vectorially, that is the direction of the DNA between two electrodes must be defined. This may be achieved by immobilizing different oligomers on top of the electrodes. We used PNA in the case of immobilization of doublestranded DNA. Several approaches have been discussed. Micropainting, i.e. the deposition of activated oligomers by a micropipette has been shown to work in principle, however, this method is very limited to the skills of the experimenter [15]. Using photochemical coupling within a focused laser beam will be an alternative approach [16] especially in cases where transparent substrates are required, e.g. for microscopic control. For immobilizing on electrodes we developed a method using an electroactive polymer to entrap the specific oligomer [17]. The chemical basis of this polymer is scopoletin. During polymerization streptavidin loaded with a biotinylated PNA-oligomer is entrapped on the surface of the active electrode. One electrode is chosen to entrap a specific oligomer, that binds to either end of the M13 DNA. Afterwards DNA is added in a low concentration and stretched to bind at both ends. The DNA is visualized by use of a fluorescence bead loaded with a third PNA-probe that binds in the center region of the M13 DNA, as is shown in Fig. 7.

FIGURE 7. Single DNA strands vectorially immobilized between IDE by polymer entrapped biotinylated PNA-oligomers. The DNA is visualized by a fluorescent bead loaded with a third PNA-probe that is specific for a sequence in the center region of the bridging M13 DNA. Both electrodes are visible as white or gray strips, resp.

CONCLUSION

Manipulation of single DNA molecules using the self-assembly feature of the DNA within a framework of electrodes has been demonstrated. Vectorial orientation of bridging DNA molecules can be visualized by use of fluorescent microbeads or nano-particles that are tagged with specific PNA-probes. The immobilization of long DNA molecules on electrode structures has been shown using an electro-polymer entrapment technique. Thus tools are now at hand to structure surfaces with DNA molecules for (bio)technological or electronic applications.

ACKNOWLEDGMENTS

Long chain PCR have been performed by Xenia Marschan, Dennie Andresen and Mandy Lorenz, kind gift of PNA-probes from Michael Bienert and Angelika Ehrlich (FMP, Berlin) is gratefully acknowledged, we also appreciate fruitful discussion with Eva Ehrentreich-Förster, Markus von Nickisch-Rosenegk and Christian Heise.

This work was supported by the German Ministry of Education and Research (BMBF) within the BioFuture framework grant no. 0311842A.

REFERENCES

1. Niemeyer, C.M., *Curr. Opin. Biotechnol.* **4**, 609-618 (2000)
2. Washizu, M., Kurosawa, O., *IEEE Trans. Ind. Appl.* **26**, 1165-1172 (1990)
3. Bier, F.F., Kleinjung, F., Ehrentreich-Förster, E., and Scheller, F.W., *BioTechniques* **27**, 752-758 (1999)
4. Nielsen, P.E., *Current Opinion Biotechnology* **12**, 16-20 (2001)
5. Hagerman, P.J., *Annu. Rev. Biophys. Biophys. Chem.* **17**, 265-286 (1988)

6. Bensimon, A., Simon, A., Chiffaudel, A., Croquette, V., Heslot, F., and Bensimon, D., *Science* **265**, 2096-2098 (1994)
7. Strick, T.R., Croquette, V., and Bensimon, D., *Proc. Nat. Acad. Sci.* **95**, 10579- 10583 (1998)
8. Kabata, H., Kurosawa, O., Arai, I., Washizu, M., Margarson, S.A., Glass, R.E., and Shimamoto, N., *Science* **262**, 1561-1563 (1994)
9. Hölzel, R., *J. Electrostatics*, in press (2002)
10. Green, N.G., Ramos, A., and Morgan, H., *J. Electrostatics* **56**, 235-254 (2002)
11. Pohl, H.A., *Dielectrophoresis*, Cambridge University Press, London, 1978
12. Green, N.G., Ramos, A., Gonzales, A., Castellanos, A., and Morgan, H., *J. Electrostatics* **53**, 71-87 (2001)
13. Hölzel, R., Gajovic-Eichelmann, N., and Bier, F.F., *Biosensors Bioelectronics*, in press (2002)
14. Bustamante, C., Smith, S.B., Liphardt, J., Smith, D., *Current Opinion Struct. Biol.* **10**, 279-285 (2000)
15. Bier, F.F. and Schmauder, R. "Nanometer addressable lateral surface structuring by use of nucleic acids" in *Coupling of Biological and Electronic Systems-2000* edited by K.-H. Hoffmann, Proceedings of the 2nd caesarium, Berlin: Springer, 2002, pp. 23- 28.
16. Heise, C., Ehrentreich-Förster, E. and Bier, F.F. unpublished results
17. Gajovic-Eichelmann, N., Ehrentreich-Förster, E., and Bier, F.F., *Biosensors & Bioelectronics* in press

NANOWIRE AND CONDUCTIVITY

DNA Nanotechnology

Masateru Taniguchi and Tomoji Kawai

The Institute of Scientific and Industrial Research, Osaka University, Mihogaoka, Osaka 567-0047, Japan

Abstract. DNA is one candidate of promising molecules for molecular electronic devices, since it has the double helix structure with π-electron bases for electron transport, the address at 0.4 nm intervals, and the self-assembly. Electrical conductivity and nanostructure of DNA and modified DNA molecules are investigated in order to research the application of DNA in nanoelectronic devices. It has been revealed that DNA is a wide-gap semiconductor in the absence of doping. The conductivity of DNA has been controlled by chemical doping, electric field doping, and photo-doping. It has found that Poly(dG)·Poly(dC) has the best conductivity and can function as a conducting nanowire. The pattern of DNA network is controlled by changing the concentration of the DNA solution.

INTRODUCTION

Inorganic solid-state devices, as represented by silicon, are based on band engineering, a technique of controlling band structure. Since it is necessary for material to contain a certain number of atoms in order to form a band structure, there is an inevitable size limit to a device to maintain bulk-like physical properties.

On the other hand, molecules have their own highest occupied molecular orbital (HOMO) and lowest unoccupied molecular orbital (LUMO) energy levels and each molecule's oxidation-reduction potential in an external electric field can be used to add and extract electrons. This idea was proposed in 1974 by Aviram and by Ratner and Carter et al., and has been studied extensively in molecular electronics in recent years [1-7]. In particular, organized and strategic studies are being promoted in the United States, headed by Defense Academy of Research Program Agency (DARPA) [8].

In this background, a DNA molecule, which carries the genetic code of all living matter, is not only important from biological, medical points of view, but is also an extremely interesting molecule as a functional material that can be utilized in nano structural applications because of its unique features.

DNA molecules have the following properties: first, the molecule contains an information and address in itself and is therefore an ultimate information material with an address of 0.34 nm pitch. Secondly, the molecule can form a wide-range of structures due to its self-organization. Thirdly, it has functions of complimentarity of base molecules and self-replication, and therefore can be copied repeatedly without error. Fourthly, it is a unique one-dimensional system from the viewpoint of low-dimensional material physics. By controlling the sequences of four kinds of bases, we

CP640, *DNA-Based Molecular Construction: International Workshop,* edited by W. Fritzsche
© 2002 American Institute of Physics 0-7354-0095-4/02/$19.00

can realize a one-dimensional superlattice with a band structure controlled in the base stacking direction, and hence can expect to observe physical properties characteristic of low-dimensional systems (e. g., Peierls transition [9]).

It is expected in particular that the first two properties can be used in the design and construction of devices based on the concept of a "programmed self-assembly", proposed by Seeman et at. [10]. The human body starts from a single cell and grows by repetition of cell divisions. In this process, eyes, heart, hands, and legs are made, based on the programming information contained in the gene, and finally, a highly functional and elaborate structural system comprising approximately sixty trillion cells appears. Following the concept of self-assembly due to these programs, creation of new material with nano-structural control and device and system constructions can be expected with the help of DNA nanotechnology.

BASIC PROPERTIES OF DNA

It is important to unveil the electric properties of DNA molecules, not only because the molecules have fundamental features as molecular devices and as nanowire, but also because DNA is an extremely important research target in the fields of molecular biology and genetic engineering, related to the mechanisms of genetic damage and mutation.

Most of the previously reported researches focus on the electric properties of the DNA where base molecules are randomly distributed. In addition, in many cases the measurement conditions, the quantities of impurity ions, and the cleaning condition of buffers were not clearly indicated, and the measurements of physical properties were performed without knowing the details of the DNA structures.

Nanostructure

We here show how effective STM is for observing the structure of DNA molecules in nanoscale. Figure 1(a) represents an STM image of the DNA double-helix structure. The structure was predicted by Watson and Crick based on X-ray analysis [11], but its clear real-space image was first obtained successfully by STM [12]. We see that short DNA such as this has a loose pitch of double helix on the surface of the solid. In addition, at the point where the double helix are relaxed, we could clearly observe the base pair constituting the DNA in double-helix structure (Fig. 1(b)) [13]. Since electric transport, DNA's electric properties and its nanostructure are all closely related, it is important to understand these relations to make devices from DNA molecules.

Electronic Structure

We investigated the electronic states near the Fermi level using photoelectron spectroscopy and X-ray absorption spectroscopy, in order to understand DNA's electric properties. The samples we used are DNA (Poly(dG)·Poly(dC) and Poly(dA)·Poly(dT) networks formed on a SiO_2 substrate [14]. The structural relaxing of the double helix, which is seen in a single DNA molecule, is expected to be limited

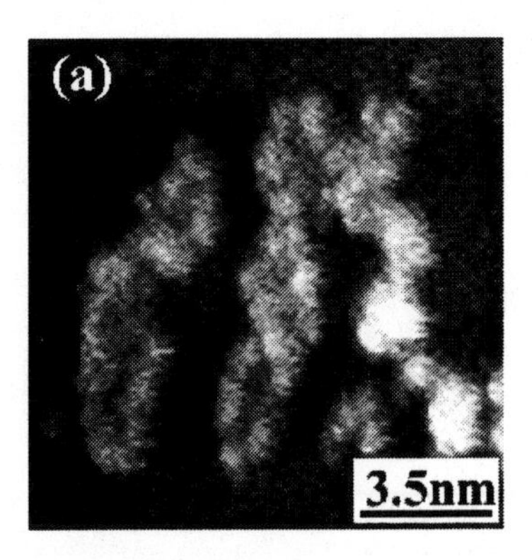

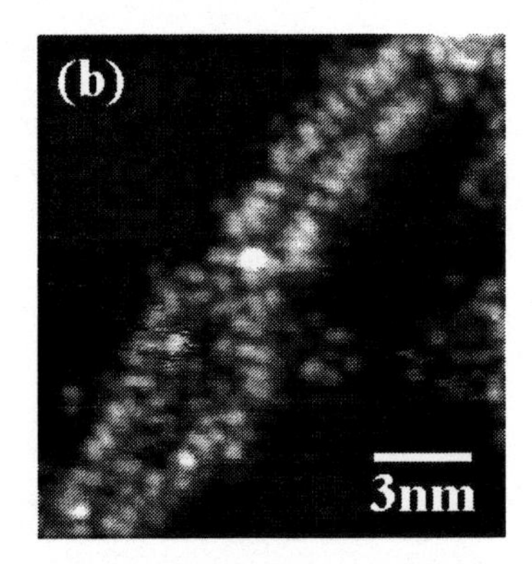

FIGURE 1. High-resolution STM images. a) A paired pAAAAAAATTTTTTT, showing double-helix structures. b) Poly(dA-dT)·Poly(dA-dT) on Cu(111) substrates.

due to the network. The experimental result is shown in Fig. 2. We found that the 1stHOMOs of Poly(dG).Poly(dC) and of Poly(dA).Poly(dT) have E-E_F = 3.6 eV, 4.3 eV respectively. Also, the unoccupied electronic states for Poly(dG).Poly(dC) as well as for Poly(dA).Poly(dT), obtained by X-ray absorption spectroscopy, have a sharp peak near E-E_F = -1 eV and a broad structure at –6 eV and over [14]. The former may be related to LUMO. Although we must be careful to interpret the unoccupied states, it is in good agreement with the measurement of the visible-ultraviolet absorption spectrum. As for the latter, the wavelengths at the maximum absorption for Poly(dG). Poly(dC) and Poly(dA).Poly(dT) are respectively given as 253-255 nm (approx. 4.9 eV) and 260-261 nm (4.8 eV), and if it corresponds to the absorption at the band edge, we can consider that the DNA molecule has an energy gap of about 5 eV. Therefore, with no doping, a DNA molecule is a wide-gap semiconductor irrespective of the kind of base.

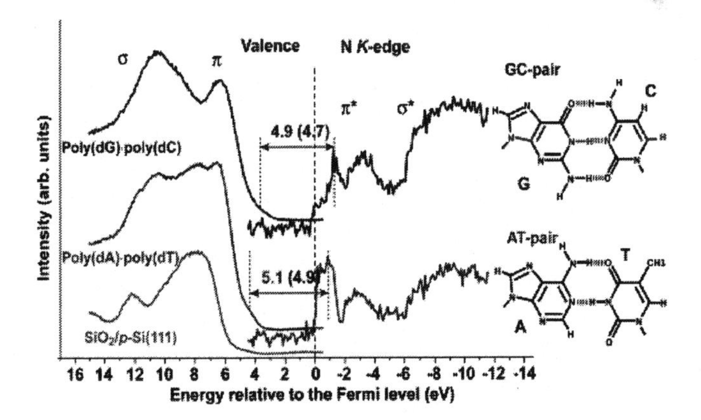

FIGURE 2. Occupied and unoccupied electronic structure near the Fermi level (E_F) of Poly(dG)·Poly(dC) and Poly(dA)·Poly(dT), absorbed onto SiO_2/p-Si(111) surfaces based on the characterization from UPS and XAS.

DNA ELECTRONICS

While DNA is an important molecule for transmitting biological genetic information, it can have a unique structure and electric properties if its structure and environment are controlled. The challenge is to make the world-smallest electric circuit using DNA. In order to perform as a real device, an electric circuit of DNA needs to have the following features: (1) circuit formation on silicon (2) switching function (3) memory function. We are trying to realize these features through various controls, e.g., modifications of DNA with molecules. We show below some of the controls.

DNA Circuit

Although it is well known that DNA has a double-helix structure, in our body it is usually entangled in proteins and is folded many times to form a ball-like structure. However, we found that under certain conditions it forms an extremely unique self-assembled network pattern on the surface of a solid in the area of more than 1 cm^2 [15]. The pattern contains 1-10 double helix chains of DNA molecules and is controllable by the combination of DNA and substrate, concentration, ionic strength, dry speed of solution, etc. We also succeeded in constructing a network on a silicon substrate, which is a necessary condition to make a device (Fig. 3(a)) [16]. The plus-charge treatment using $MgCl_2$ has found to be effective for the DNA adsorption on SiO_2.

Applying a chemical treatment on the surface of Si to produce hydrophobic Si-H and hydrophilic SiO_2 and forming a DNA self-assembled network circuit only on the SiO_2, we showed that the patterning of a DNA circuit by the conventional etching technique is possible (Fig. 3(b)). Inside DNA there are bases, which carry genetic information, forming π-stacks, and hydrophilic phosphate groups lie outside. Since phosphate groups are electrically insulating, a DNA molecule is like a conducting wire with an insulating coating.

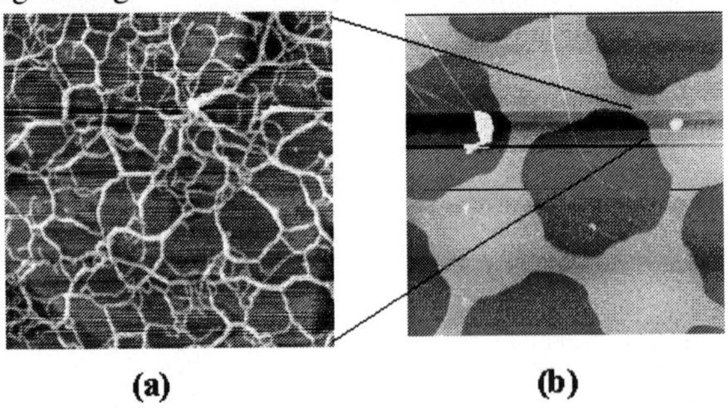

(a) (b)

FIGURE 3. AMF images of a) DNA network on a SiO_2/Si(111) substrate and b) the surfaces of Si-H and SiO_2.

We extended and fixed a DNA between nano-size metal electrodes formed on a SiO_2 substrate, aiming to clarify the nano-scale physical properties and integration (Fig 4). In this method, developed by Washizu et. al. [17], a high-frequency electric

Field (1 MHz, 1 MV/m) is applied between the two electrodes, which extends DNA molecule chains due to the dielectric extension, and the DNA connects via a phosphate group to the metal oxides (e. g., Al_2O_3, TiO_2) formed on the surface of the metal.

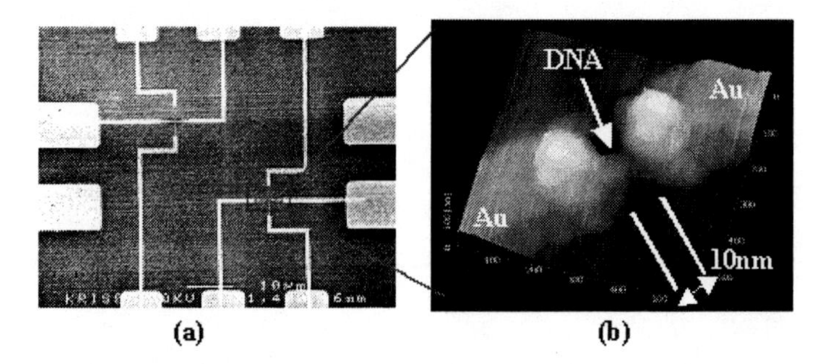

FIGURE 4. a) A SEM image of an Au/Ti nanoelectrode of which spacing is about 10 nm. b) An AFM image of DNA molecules trapped between two electrodes.

DNA-FET

Yoo et. al. extended and fixed a DNA bundle between electrodes separated by about 20 nm using the same kind of device structure as shown in Figure 4(b) and measured the temperature dependence of the electric properties [18]. The electric current was found to be proportional to sinh(0.67 V) and the conduction mechanism of small polarons where the lattice deformation and the conduction electrons are coupled is indicated based on this result. Various models have now been proposed as conduction mechanisms in DNA molecules, including hopping conduction, tunneling and superexchange between DNA chains, and polaron conduction [19-22]. Investigation of physical properties using these nano-electrodes, including confirmation of reproducibility and control of impurities such as counter ions, which exist for stabilization of a DNA molecule, is very effective for understanding fundamental features of new nanodevices.

Applying an additional electric field from the back of the Si substrate and measuring the FET property of the back-gate type, we obtained very interesting results. In the FET based on Poly(dA)·Poly(dT), application of positive bias on the gate electrode results in an increase of the drain current, which indicates n-type behavior (Fig. 5(a)). On the other hand, DNA-FET based on Poly(dG)·Poly(dC) shows the p-type electric property (Fig. 5(b)) [18]. This is reflected by the fact that among the four bases, guanine has the lowest ionization potential [23, 24]. DNA is essentially an insulator, but if the transistor control by the combination of base sequence and electric field is possible, it is not a dream to realize FET of an ultimately small size. In this way, we can expect creation of new functional devices, combining functional material and silicon by the nano-scale control of interface and structure.

The synthesized DNA we used this time is a self-assembled network comprising 50 base pairs and there is a nick (structural defects) at every 50 base pairs. In the fine structure of DNA observed by the Non-contact(NC)-AFM measurement, there is a

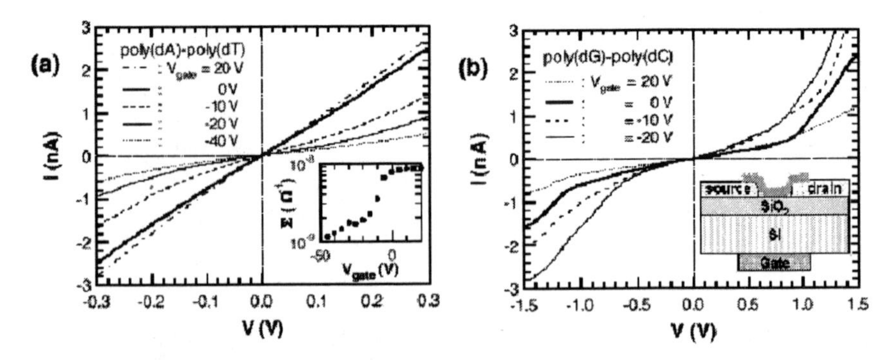

FIGURE 5. The *I-V* curves measured at room temperature for various values of the gate voltage (V_{gate}) for a) poly(dA)-poly(dT) and b) poly(dG)-poly(dC). In the inset of a), the conductance at $V = 0$ is plotted as a function of V_{gate} for poly(dA)-poly(dT). The inset of b) is the schematic diagram of electrode arrangement for gate dependent transport experiments. DNA molecules are trapped between two nanoelectrodes of which spacing is about 20 nm.

low-height part at about 20 nm intervals [25]. Accordingly, we need to take into account the electron hopping between the nicks as an influential factor to the electric properties. Recently, we succeeded in synthesizing a damage-free DNA repaired by ligation using enzyme. Studies on this damage-free DNA are in progress.

Chemical Doping

DNA essentially does not absorb visible light and therefore is an insulator with a 5 eV band gap. We tried to dope iodine to DNA molecules. We left DNA molecules in a vacuum of 5×10^{-5}Torr more than 8 hours and then iodine gas was introduced at room temperature. We then found that the electric resistivity of Poly(dG) . Poly(dC) decreased more than three orders of magnitude. Figure 6 shows the change of the electric resistivity of a DNA membrane measured with Au electrodes of 100 nm gap. We can see the resistivity drops from 10^{11}. to 3×10^{7}. by the I_2 doping. Just after the doping, the absorption at 3.6 eV, which is characteristic to I_3, is clearly observed, but the peak gradually disappears as time passes, although the resistivity of the DNA membrane remains low. However, the ultraviolet absorption peak, which indicates the inter-band transition (...), shifts its position from 4.6 eV to 4.9 eV, as time passed after the I_2 doping. In this way, DNA molecules are expected to be a semiconductor with a wide gap, whose electric properties are controllable by doping.

Photo-doping

One of the features of DNA is that many molecular modifications with various functionalities are already known, and therefore an optical conduction property is expected to be accompanied as part of the integration.

Usually, acridine orange (AO) molecules are intercalated into a base stack and are used as a fluorescent marker. This time we mixed dG-dC and AO and formed an AO-

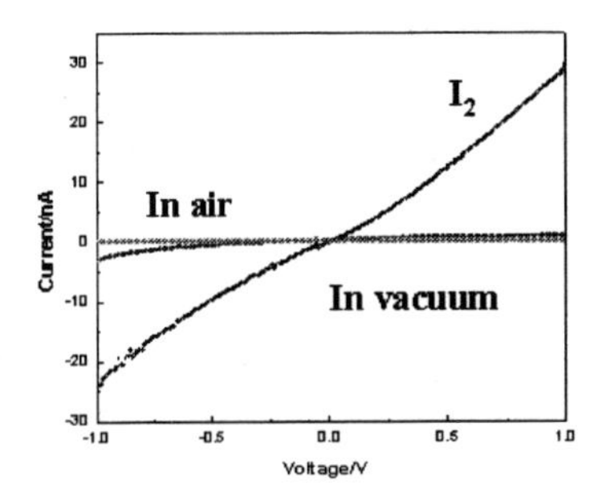

FIGURE 6. *I-V* curves for Poly(dG)·Poly(dC) doped with iodine molecules.

DNA network on a mica [26]. In our experimental conditions, AO is intercalated at about every 10 base pairs. For solvent, we used 10 mM Tris-HCl, pH 7.6 solution.

We measured the *I-V* characteristics of conductive AFM tip [27]-DNA with electrodes separated by 50-100 nm under an Xe lamp (12 mWcm^{-2}). In order to prevent inter-band transitions, we used an optical filter to cut out wavelengths under 340 nm. When the sample bias was plus for the Au-tip, increase of photoelectric current was observed (Fig. 7(a)). As a reference experiment we did the same thing on a DNA network with no AO doping and found no increase of electric current due to the beam irradiation. In addition, for the DNA (dA-dT)comprising adenine and thymine only, the resistivity change due to the beam irradiation is around one third or one fourth compared to dG-dC. Although the excited electrons from AO are generally considered not to be quenched by base, we clearly observed the improvement of the optical conduction in the DNA comprising G-C.

We also observed the wavelength dependence of the beam-irradiation effect and found that it responds noticeably to the light of wavelength about 500 nm, which coincides with the wavelength of the optical absorption peak of AO-DNA. Therefore,

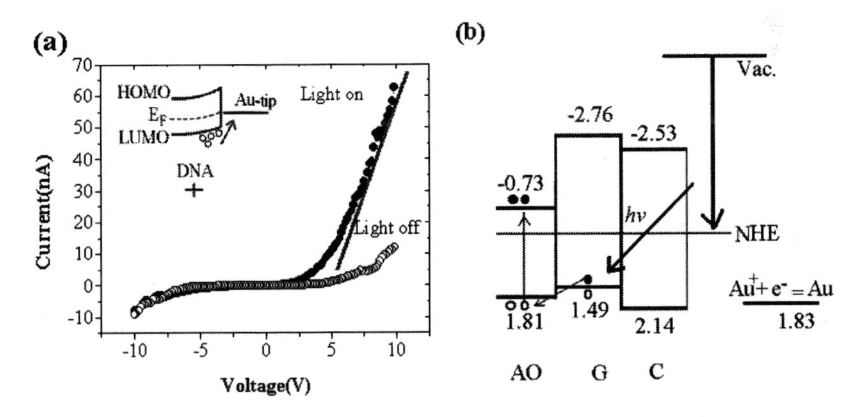

FIGURE 7. a) *I-V* curves for the AO-DNA network with and without light irradiation. Inset: Energy diagram of AO-DNA and the Au AFM tip. b) Schematic diagram of the redox potentials.

we see the increasing carriers due to the optical excitation would affect the electric resistivity. The emitted light is absorbed in AO and electron-hole pairs are created. Then electrons are transferred from guanine and the hole number in DNA increases, which leads to a larger electric current (Fig. 7(b)). The same tendency is observed in the experiment on a similar DNA thin-film crystal. Although the resistivity measured in a vacuum is about ten times as large as in the air, the wavelength dependences are the same [26].

CONCLUSION

In this review, basic characteristics of DNA and electrical properties of AO-DNA network, several DNA molecules and DNA thin films have been described. Furthermore, DNA patterning onto silicon has been referred, because the fusion with conventional electronic devices based on silicon semiconductors is the central problem in the realization of novel molecular devices. Since DNA is the molecule that has the ability of "Programmed Self-assembly", the combination with DNA wire, various protein molecules and nano-particles together with the nano-lithography method will give an opportunity to make an advanced nanoscale electronic circuit, such as brain mimetic devices that automatically form, repair and switch electrical circuits with appropriate programming in the DNA.

ACKNOWLEDGMENTS

We would like to thank our laboratory and all other collaborators. We are especially grateful to Prof. Yoo of the Yonsei University for preparing nanogaps and FETs, and to Prof. Shin of the Institute for Solid State Physics of the University of Tokyo, Fujimori of the Department of Physics of the University of Tokyo and M. Kawai of the Institute of Physical and Chemical Research for photoelectron spectroscopy.

REFERENCES

1. Aviram, A., and Ratner, M., *Chem. Phys. Lett.* **29**, 277-283 (1974).
2. Heath, J. R., Kuekes, P. J., Snider, G. S., and Williams, R. S., *Science* **280**, 1716-1721 (1998).
3. Joachim, C., Gimzewski, J. K., and Aviram, A., *Nature* **408**, 541-548 (2000).
4. Wada, Y., Tsukada, M., Fujihira, M., Matsushige, K., Ogawa, T., Haga, M., and Tanaka, S., *Jpn. J. Appl. Phys.* **39**, 3835-3849 (2000).
5. Niemeyer, C. M., *Appl. Phys. A.* **68**, 119-124 (1999).
6. Reed, M. A., and Tour, J. M., *Scientifc American*, Scientific American, New York, 2000, p69-pp76.
7. Reed, M. A., Zhou, C., Muller, C. J., Burgin, T. P., and Tour, J. M., *Science* **278**, 252-254 (1997).
8. Kwan, S., Materials Today,
9. Peierls, R. E., *Quantum Theory of Solids*, Oxford Univ. Press, London, 1955.
10. Winfree, E., Liu, F., Wenzler, L. A., and Seeman, N. C., *Nature* **394**, 539-544 (1998).
11. Watson, J. D., and Click, F. H. C., *Nature* **171**, 737-738 (1953).
12. Tanaka, H., Hamai, C., Kanno, T, and Kawai, T., *Surf. Sci. Lett.* 432, L611-L616 (1999).

13. Tanaka, H., and Kawai, T, *Biophysics* **231**, 336- (2000).
14. Furukawa, M., Takeuchi, T., Tanaka, S., Kato, H. K., Komeda, T., Kawai, M., Kawai, T., and Shin, S., *Phys. Rev. B* in press.
15. Kanno, T., Tanaka, H., Miyoshi, N., and Kawai, T., *Jpn. J. Appl. Phys.* **39**, L269-L270 (2000).
16. Tanaka, S., Cai, L. –T, Tabata, H., and Kawai, T., *Jpn. J. Appl. Phys.* **40**, L407-L409 (2001).
17. Washizu, M., Kurosawa, O., Arai, I., Suzuki, S., and Shimamoto, N., *IEEE Trans. Ind. Appl.* **31**, 447-456 (1995).
18. Yoo, K. –H., Ha, D. H., Lee, J. –O., Park, J. W., Kim, J., Kim, J. J., Lee, H. –Y., Kawai, T., and Choi, H. Y., *Phys. Rev. Lett.* **87**, 198102-198105 (2001).
19. Jortner, J., Bixon, M., Langenbacher, T., and Michel-Beyerle, M. E., *Proc. Natl. Acad. Sci.* **95**, 12759-12765 (1998).
20. Bixon, M., Giese, B., Wessely, S., Langenbacher, T., Michel-Beyerle, M. E., and Jortner, J., *Proc. Natl. Acad. Sci.* **96**, 11713-11716 (1999).
21. Giese, B., Wessely, S., Spormann, M., Lindemann, U., Meggers, E., and Michel-Beyerle, M. E., *Angewandte Chem. Int. Ed.* **38**, 996-998 (1999).
22. Shuster, G. B., *Acc. Chem. Res.* **33**, 253-260 (2000).
23. Siedel, C. A. M., Schulz, A., and Sauer, M. H. M., *J. Phys. Chem.*, **100**, 5541-5553 (1996).
24. Steenken, S., and Jovanovic, S. V., *J. Am. Chem. Soc.* **119**, 617-618 (1997).
25. Tabata, H., and Kawai, T, *J. Jpn. Assoc. Cryst. Growth* **28**, 56-64 (2001).
26. Gu, J., Tanaka, S., Otsuka, Y., Tabata, H., and Kawai, T., *Appl. Phys. Lett.* **80**, 688-690 (2002).
27. Cai, L., Tabata, H., and Kawai, T., *Appl. Phys. Lett.* **77**, 3105-3106 (2000).

A Construction Scheme For A SET Device Based On Self-Assembly Of DNA And Nanoparticles

Wolfgang Fritzsche, Gunter Maubach, Detlef Born, J. Michael Köhler, Andrea Csáki

Biotechnical Microsystems Department
Institute for Physical High Technology
Jena, Germany
fritzsche@ipht-jena.de

Abstract. The unique potential of molecular nanotechnology is based on the fabrication of materials and devices starting from molecular units. Comparable to and based on the synthetic approach in supramolecular chemistry or molecular biology, an extended toolbox of molecular units as well as tailored reactions is provided by these fields. On the other hand, the progress in synthesis of molecular structures is not directly transferable into technical applications, what is mainly due to a missing integration of the synthetic products into technological interfaces and environments. Self-organization as used by nature to create complex organisms appears to be a solution to this dilemma. We propose a scheme for the realization of a single electron-tunneling (SET) device based on this principle, and demonstrate the realization of various steps toward this aim, especially a technique for immobilizing exactly one DNA molecule in a microelectrode setup based on self-assembly.

INTRODUCTION

The traditional approach in the ongoing miniaturization, especially in microelectronics, is widely based on the top-down approach. It includes the fabrication of small features starting from larger structures, as described in Fig. 1 for a typical process in planar microfabrication. A layer of the material for microstructuring is deposited on a substrate and covered by a photoresist (Fig. 1a). The geometry of the feature is exposed into the resist (b), developed and transferred into the desired material (c). An important advantage of this approach is the compatibility with standard technologies, not only relating to required equipment but also in the case where interfaces, contacts etc. for the microfabricated structure are needed (d). This advantage still predominates the growing difficulties with ongoing miniaturization, when the physical limits (e.g. optical resolution or grain size of used materials) seem to limit further development.

An alternative for the fabrication of small devices and materials is the "bottom-up" approach (Fig. 2). It is mainly based on molecular units (a), which are assembled into larger structures (b). This idea is motivated by the progress in fields like synthetic

CP640, *DNA-Based Molecular Construction: International Workshop,* edited by W. Fritzsche
© 2002 American Institute of Physics 0-7354-0095-4/02/$19.00

chemistry or DNA technology, where highly complex molecules can be constructed in a defined way. However, a technical application of these structures is often not possible, because they are not embedded in a technical interface. To realize this connection with molecular structures, a variety of methods was developed, which mostly miss the needed control to achieve a high efficiency required for a technical application. Another disadvantage is the serial character of the approaches, which hampers the needed high throughput for the envisioned development towards a complementary or even superior microfabrication technology. These problems could be overcome by the use of self-assembly techniques known from biology and chemistry, with the special emphasis on the application of DNA [1].

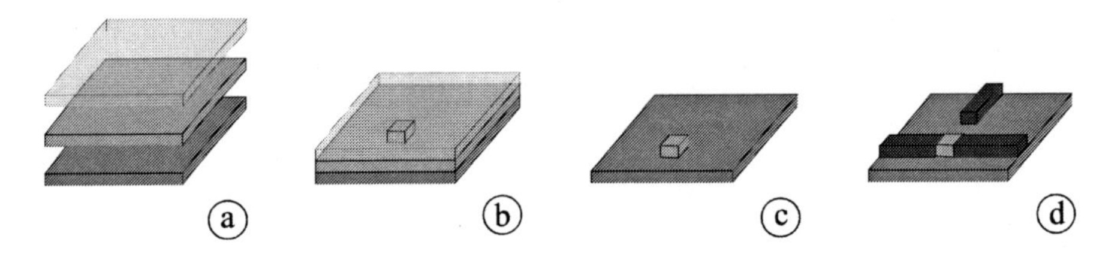

FIGURE 1. Top-Down approach as standard technique for further miniaturization today. The scheme illustrates the step of microlithography, starting with the deposition of thin layers of the desired material on a substrate and the coating with a light (or electron) -sensitive resist layer (a). The resist is structured by exposure of a light (electron) pattern (b), and developed (c). The resulting structure in the resist is transferred into the desired material (d); the realized structure can be embedded into interface structures using similar steps.

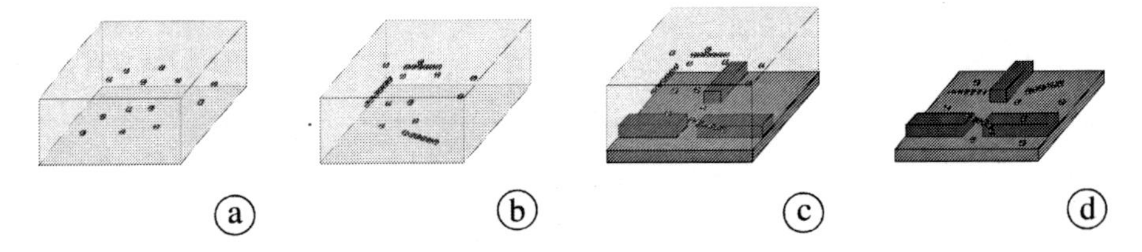

FIGURE 2. Scheme of Molecular Nanotechnology as a Bottom-Up approach. Starting with molecular units (a), an assembly step is used to create materials or devices (b). The assembled structures have to be interfaced, e.g. by incubation of substrates (c) prior to drying (d).

MATERIALS AND METHODS

Chip Fabrication And Modification

The electrode structures were microfabricated on a thermally oxidized silicon wafer by standard photolithographic techniques. A 100 nm gold layer was deposited by sputtering, followed by a lift-off process. After microfabrication, the whole wafer was incubated with 1 mM octadecyl-trichloro-silane in freshly distilled toluene overnight at room temperature. Then it was washed twice with dry toluene and once with ethanol, followed by excessive washing with pure water. Afterwards, the chips were immersed in a solution of 2 mM cysteamine (2-aminoethanethiol, Sigma Aldrich) in pure ethanol for at least 1 hour, prior to drying in a nitrogen flow.

Flow Chamber And DNA Incubation

A microfluidic chip chamber (IBA Heiligenstadt, Germany; Fig. 8) was utilized to apply a laminar flow to the inserted chip. The lid is structured to give a proper orientation of the chip. The flow through the camber is realized by a pump (Ismatec, Wertheim) and was adjusted to 1 ml/min. 250 ng/ml λ-DNA (MBI Ferment St. Leon-Rot, Germany) in 10 mM Tris/HCl pH 8.0 was pumped in a closed loop through the chamber for 30 minutes. After removal from the chamber, the chip was dried in a nitrogen flow.

Imaging

SFM images were taken with a Dimension 3100 (Digital Instruments, Santa Barbara) using tapping mode in air. SEM imaging was with a digital scanning electron microscope DSM 960 (Zeiss, Germany).

RESULTS AND DISCUSSION

Molecular nanotechnology attracts a lot of visionary thoughts on one hand. On the other hand, the last years also demonstrated a tremendous amount of experimental developments towards the characterization and manipulation on the molecular scale, as an important requirement for a realization of the envisioned potential. Two important points have to be addressed on the way from interesting experiments to an established technology: The seamless insertion of the molecular structure in today's technical environment (e.g., electrical wiring), and the potential for a highly paralleled fabrication method.

Our goal is to demonstrate an approach that addresses both points by relying on self-assembly processes. We chose a Single Electron Tunneling (SET) transistor as demonstrator device. This kind of transistor promises to reach the ultimate limit in signal generation, the single electron. This sensitivity could minimize energy consumption, enabling therefore a higher integration because of the decrease in heat dissipation due to minimized electron flow through the microstructured integrated circuits. It could provide an example for the potential of molecular nanotechnology, because the applicability of a SET device is mainly influenced by its size.

The typical setup of an SET transistor is given in Fig. 3. An electron confinement is surrounded by tunneling barriers, which separate it from the contacts. A gate electrode is situated in the vicinity (up to several hundreds of nanometers) of the confinement. The working temperature of a SET transistor is influenced by the capacity of the confinement, which in turn is determined (beside other parameters) by the size. So confinements in the low nanometer range could work at room temperature, in contrast to the needed low temperatures for today's standard microfabricated SET transistor devices.

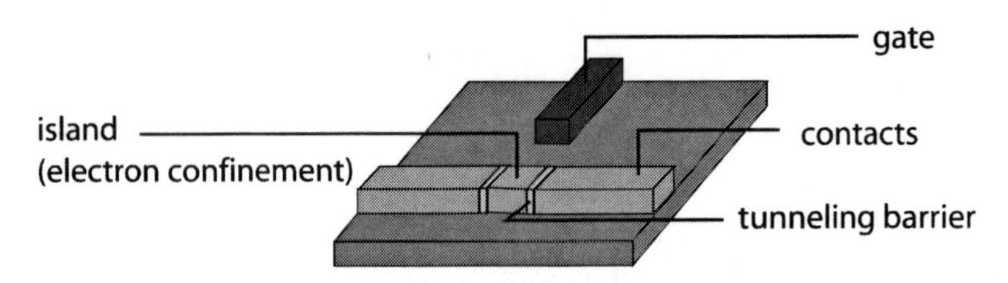

FIGURE 3. Scheme of a SET transistor realized by conventional fabrication technology

The development in this field is determined by the search for improvement of or novel developments for the microfabrication technology needed to prepare small electron confinements in a time and cost effective way. Beside the modification of traditional microstructuring techniques following the top-down approach, novel approaches based on the use of metal nanoparticles were proposed [2]. Metal nanoparticles are colloidal metal particles with dimensions in the range between approximately 0.8 and up to 250 nm; their interesting optical and electrical properties were subject of numerous studies. They have been used as marker in light and electron microscopy [3, 4], and recently their application for DNA chip technology was demonstrated [5]. The application of these nanoparticles for the electron confinement is mainly based on the highly controllable diameter. The particles provide metal structures of a defined geometry and size in the lower nanometer range with a sub-nanometer precision. Beside the size of the confinement, the properties of the tunneling barrier are important. In the case of metal (e.g. gold) particles, this barrier can be realized using molecular layer which self-organize on the particle surface. So a tailor-made coating with tunable electronic properties is in principle possible.

The use of nanoparticles as confinement was demonstrated in experiments based on measurements of nanoparticles in ordered monolayer arrays [2] or as individual structures [6]. These experiments revealed the difficulty with this approach regarding a defined wiring of the particle. There are several ways to solve this problem. One group of experiments used scanning probe techniques (STM or conductive SFM) to contact the particles, the second contact is usually realized by the substrate (Fig. 4a). Although scanning probes allow the wiring of a variety of designs and patterns, it is a slow serial approach with complicated equipment and the contact resistance is not constant (depending on force, oxidation etc), thereby preventing a technical application.

A preferred setup would include only microstructured electrodes, so that the geometry is fixed and the fabrication is simplified. In its simplest form, a three-electrode setup would be used for the random adsorption of nanoparticles (Fig. 4b). Without a driving force in favor of adsorption in the electrode gap, this approach misses the needed efficiency. An interesting technique applies electrostatic trapping to position a particle in the gap (Fig. 4c, [7]). Although this method fulfills the requirement for a defined and potentially parallel technique, it could not be established as a standard method, which is probably due to the required complicated nanostructures.

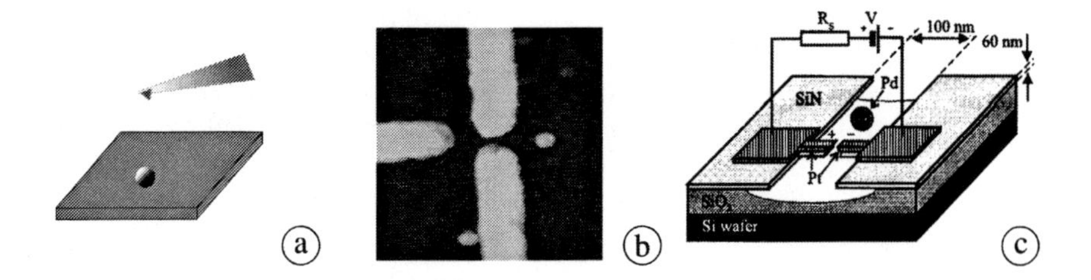

FIGURE 4. Various implementation of nanoparticle as electron confinements. a) After random adsorption of nanoparticles on a substrate, scanning probe techniques are used for wiring. b) Random adsorption onto microstructured three electrode structures (Csaki, unpublished results). c) Electrostatic trapping of one particle in a nanogap [7].

So what is needed for an ideal technology? Based on simple steps, a nanoparticle should be somehow guided into the right position, and this process should be easily paralleled and should require only standard steps of microfabrication. The guidance could be realized by self-organization, for example by the use of pairs of complementary molecules. Known examples for such pairs are thiol-gold, biotin/(strept)avidin, and DNA with complementary sequences. The use of DNA to bind nanoparticles on specifically DNA-modified surface regions was already described [8, 9], and is the foundation for the application of nanoparticles as marker in DNA-chip detection [5]. It is ideal for parallelization because of its large sequence variability, but also due to the highly developed technology for DNA manipulation.

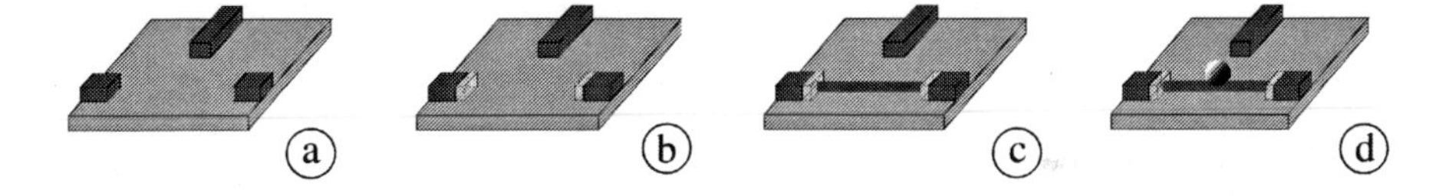

FIGURE 5. Setup of a SET device by means of molecular nanotechnology. A three-electrode structure is microfabricated (a), and the ends of two contacts are modified (b) to enable binding of a long DNA molecule in a stretched conformation between both electrodes (c). The immobilized DNA serves as a tool to position a DNA-modified nanoparticle in the gap (d).

Immobilizing Of Individual DNA Molecules

Self-assembly would solve the problem of specific binding, but with the result of numerous immobilized particles even for binding areas in the lower micrometer region. Another problem is the precision of the positioning effect: The particle has to touch both contacts. This point is not addressable by standard microfabrication techniques. A solution to this problem would be the application of long DNA as positioning framework. Compared to short DNA molecules, which pack very dense on surfaces (e.g., on DNA chips), longer DNA (micrometer range) exists usually in a globular state in solution or on surfaces. These coils exclude other molecules by electrostatic repulsion, so that the footprint can be rather large (approaching the square micrometer range) in the case of longer DNA (for example lambda DNA with a length of 16 µm). So it should be possible to bind only one long DNA molecule on a binding area, when the area size is smaller than the footprint of the DNA. Using this principle, one could position one DNA molecule between two contacts (Fig. 5a-c). Therefore, the ends of the contacts were used as binding areas by modification to induce specific binding (Fig. 5b). The small size of the binding area limits the number of immobilized DNA molecules to one, when long lambda DNA (16 µm length) is used (Fig. 5c).

Another point to address is the coiled state of long DNA in solution. This coil has to be extended after the binding of one end to expose the free end to the second binding area, which is situated in a distance approaching the contour length of the molecule. To induce a stretching, several techniques are described, ranging from electric fields [10, 11] over magnetic beads [12] to laminar flow [13].

Positioning Along A DNA Strand

As a result of the proposed steps, an individual DNA molecule is positioned in the electrode gap. Based on the in situ hybridization techniques, several authors proposed to use the base sequence of DNA molecules for addressing a specified position along the strand by a complementary DNA (cf. [11]). The realization depends on the kind of the immobilized DNA molecule. Single-stranded DNA allows free access for the complementary molecule, but the manipulation is hampered by the high flexibility, which results in a highly coiled confirmation in an undisturbed state. When double-stranded DNA is used, the overall stiffness of the molecules is increased, but the access for a single-stranded complementary DNA is not straightforward. However, based on the large body of experiences from in situ hybridization and other DNA labeling experiments from microscopy, a variety of techniques are known to solve this problem. In our case, the single-stranded DNA complementary to a specified region of the immobilized long DNA would be labeled with a gold nanoparticle, so that this particle is positioned in the electrode gap (Fig. 5d).

Fabrication Of Microstructures For DNA Binding

A first step in the realization of the scheme described above is the microfabrication of the binding structures. They should meet geometrical, biochemical and electrical requirements. The geometry is determined by the need for small binding areas (in the order of the footprint) and the preferred gap size. Biochemical aspects are the modification of the binding areas, but also a passivation of the gap to minimize unspecific binding. Because the aim of the setup is an electrical device, the wiring has to be included, and an isolating substrate is needed. An additional point are the possible application of an electric field to induce stretching, so needed additional electrode should be included.

The realized structured is shown in Fig. 6. On a substrate covered with silicon oxide, gold electrode structures were microfabricated using standard photolithographic procedures. An overview in Fig. 6a illustrates the setup of four gaps. Figure 6b shows a zoomed perspective view of the details of one gap; the two electrode structures and the additional third electrode (left) for application of an electrical field for stretching purposes are visible.

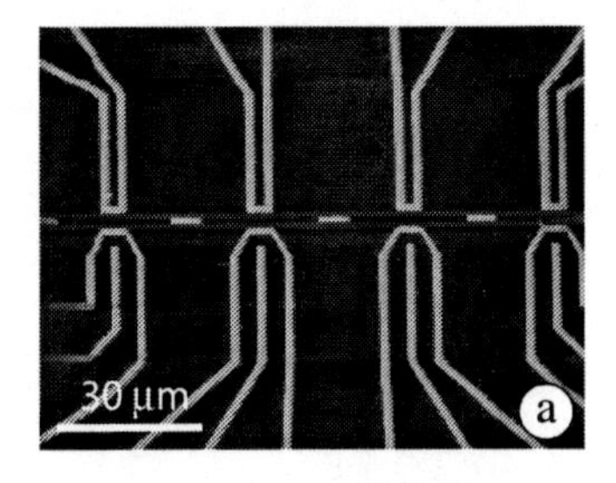

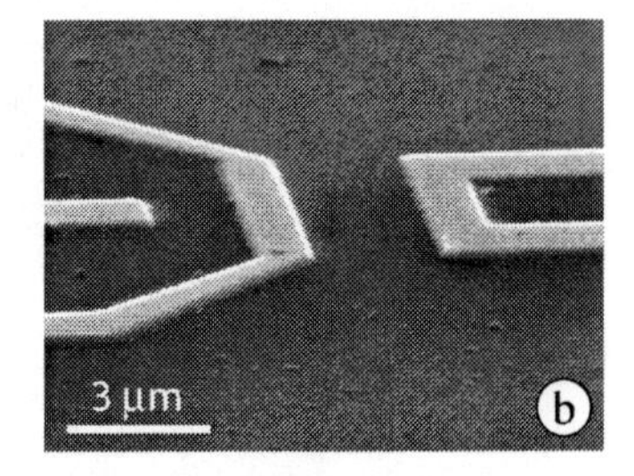

FIGURE 6. Electrode structure for binding of DNA. a) SFM overview showing four electrode gaps. b) SEM image with a magnified perspective view of one gap.

Chemical Surface Modification For Defined DNA Binding

To utilize self-organization for the binding of the DNA, binding areas with a high affinity to the ends of the DNA molecule are needed, together with a surrounding background with minimal affinity to the molecules of interest. High affinity binding areas for lambda-DNA based on immobilized DNA oligonucleotides with a sequence complementary to the overlapping single-stranded cos-ends were demonstrated [14, 15]. In the presented approach, binding is based on electrostatic interactions of positively charged microstructures and the negatively charged DNA. The well-described thiol-gold affinity was applied to charge the gold structures on the surface by incubation the complete chip with cysteamine. A coverage of the gold electrodes due to the binding of the sulfur-containing compound is observed [16], resulting in the desired surface charge induced by the immobilized amine groups.

Unspecific binding is a general problem when working with molecules on or near surfaces, as in microfluidics, biochip technology, or substrate-based molecular nanotechnology. This effect hampers directed surface immobilization by molecule binding outside the desired areas. A procedure based on silanization with octadecyl-trichloro-silane [17] was utilized to minimize DNA binding outside of the electrode structures. This modification results in a surface passivated against adsorption of DNA. The extent of passivation was tested using short DNA labeled with nanoparticles [18]. A substrate with electrodes on a silicon oxide background was silanized, and the gold electrode was modified with capture DNA for control purposes. This substrate was then incubated with complementary nanoparticle-labeled DNA. SFM imaging of the resulting surface clearly reveals nanoparticle binding on the electrodes (Fig. 7, top). However, the passivated background shows significantly fewer particles over the range of several square micrometers (Fig. 7, bottom). This result demonstrates the high efficiency of surface passivation against DNA adsorption using octadecyl-trichloro-silane.

FIGURE 7. Demonstration of the efficiency of surface passivation. A substrate surface (bottom) was passivated against unspecific DNA binding by silanization with octadecyl-trichloro-silane. For testing, the electrode (top) was modified with DNA. Incubation of the whole substrate with nanoparticle-labeled and complementary DNA resulted in a high binding on the electrode (top), but only a low number of labeled DNA on the substrate (bottom).

Immobilization Of The DNA In The Electrode Gap

An extended conformation of the immobilized DNA is required for a defined topological setup. So after binding of one end onto the first electrode, the second end has to be presented to the remaining electrode in order to facilitate binding. As already discussed, there are several possibilities to extend the coiled conformation of the DNA. Although laminar flow of a liquid is a simple approach, the difficulty lies in the control of the parameters. We utilize a liquid chamber, where the flow over the chip surface is well defined (Fig. 8). This is achieved by a channel cavity between chip surface and the bottom part of the liquid chamber, where the incubation solution containing the lambda-DNA flows (Fig. 8a). The liquid chamber can hold two chips, which can be incubated in parallel using separate in- and outlets (Fig. 8b).

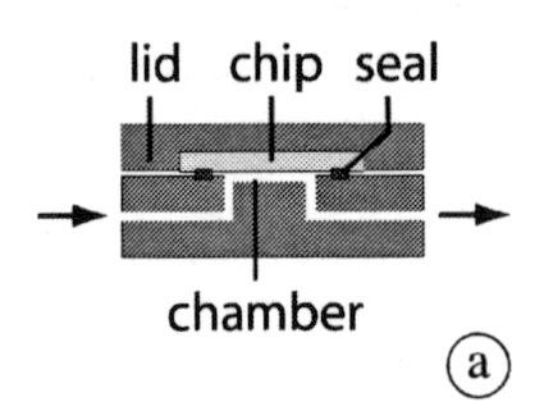

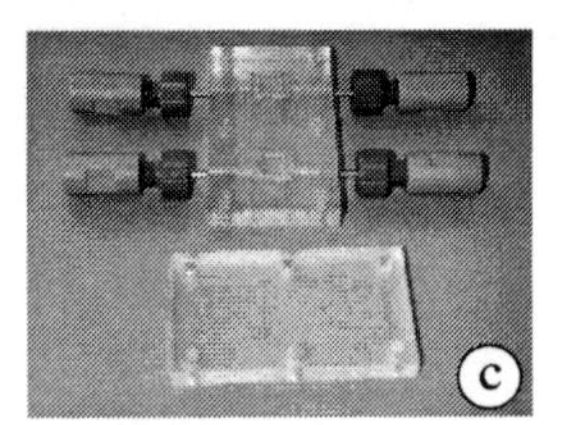

FIGURE 8. Microfluidic chamber for realization of a laminar flow of DNA solution over the chip surface in order to realize stretched molecules. a) Scheme of the chamber, the chip is inserted up side down (a). Overview of the chamber in a closed position (bottom view, b) and opened (c).

The efficiency of flow stretching was tested using lambda-DNA adsorbed onto mica [19]. Samples incubated with DNA in a laminar flow using the chamber showed clearly a stretched conformation of the molecules oriented along the flow direction, control experiments without flow resulted in coiled molecules without orientation.

After testing the flow chamber successfully, experiments with the surface-modified microstructured electrode substrates and lambda DNA were conducted. The success of binding was evaluated using SFM (Fig. 9). Besides electrode gaps with no DNA, electrode gaps with immobilized DNA showed only one individual molecule (arrows) spanning the distance between the electrodes. The microscopical images resolved the molecules clearly enough to identify them as an individual, continuous double-strand. However, the ends of the molecules could not be clearly resolved, so that the connection of the DNA to the electrodes cannot be characterized based on imaging, as discussed in more detail elsewhere [19]. Although the efficiency of bridged gaps was with 10-30% still low, the restriction in every successful case to exactly one molecule per gap appears very promising. It is a confirmation of the applied approach based on a small ratio of binding area to footprint to reach a limitation to a single molecule.

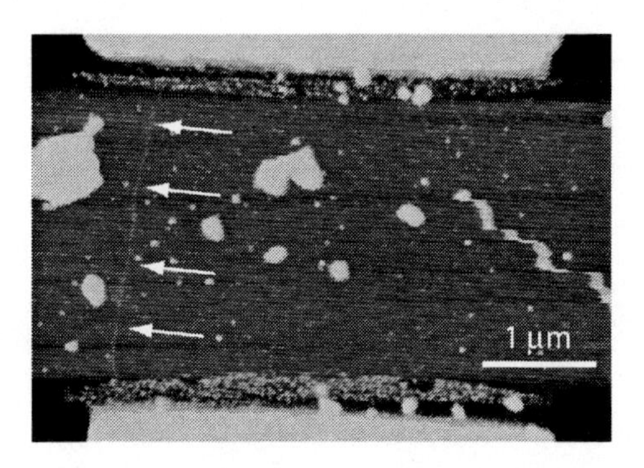

FIGURE 9. Individual DNA molecule (arrows) stretched between two microelectrodes

Outlook

After solving the key step of immobilization of exactly one DNA molecule in the electrode gap, the positioning and the wiring of a nanoparticle remain as task for further experimental developments. As mentioned before, the positioning could be based on in situ hybridization of a complementary strand with the nanoparticle along the immobilized DNA. Using PNA, a peptide-backbone analogue of DNA, the efficiency of this process could be significantly enhanced. Another line of experiments is aimed at the application of triple DNA constructs for positioning ([20], Schütz et al. unpublished results), or the construction of a T-shaped insert created from three different designed oligonucleotides fitted into the long DNA by enzymatic restriction, which binds the nanoparticle on its free end (Maubach, unpublished results).

The nanoparticle-positioning is closely connected with the final step in SET device fabrication, the wiring of the electron confinement. Although the electrical conductivity is still a matter of debate, it is not yet fully understand, so that one has to look for alternatives. We think that the specific deposition of conductive material along the DNA is a promising route to provide an electrical connection between the microstructured electrodes and the nanoparticle [14, 21-23].

So although the described approach presents work in progress, key problems as the restriction onto an individual molecule are already solved or are addressable based on introduced methods as e.g. the DNA metallization. Therefore, the construction scheme for a SET transistor using a nanoparticle as electron confinement and utilizing a single DNA molecule for defined positioning is a realistic approach toward a convincing demonstrator for the potential of molecular nanotechnology.

ACKNOWLEDGMENTS

We would like to acknowledge the assistance with sample preparation by K. Kandera and H. Porwol, help with REM characterization by F. Jahn, and the open and valuable discussions with M. Mertig and R. Seidel.

This work was supported by the DFG (FR 1348/3-4) and the Volkswagen Foundation (Priority Area: Physics, Chemistry and Biology with Single Molecules).

REFERENCES

1. N. C. Seeman, Nanotechnology **2**, 149 (1991).
2. R. P. Andres, T. Bein, M. Dorogi, S. Feng, J. I. Henderson, C. P. Kubiak, W. Mahoney, R. G. Osifchin, and R. Reifenberger, Science **272**, 1323 (1996).
3. G. M. Hodges, J. Southgate, and E. C. Toulson, **1**, 301 (1987).
4. M. Horisberger, Scanning Electron Microscopy **II**, 9 (1981).
5. W. Fritzsche, Reviews in Molecular Biotechnology **82**, 37 (2001).
6. D. Davidovic and M. Tinkham, Applied Physics Letters **73**, 3959 (1998).
7. Bezryadin, C. Dekker, and G. Schmid, Applied Physics Letters **71**, 1273 (1997).
8. M. Niemeyer, B. Ceyhan, S. Gao, L. Chi, S. Peschel, and U. Simon, Colloid and Polymers Sciences **279**, 68 (2001).
9. Csáki, R. Möller, W. Straube, J. M. Köhler, and W. Fritzsche, Nucleic Acids Research **29**, e81 (2001).
10. M. Washizu, this volume (2002).
11. F. F. Bier, N. Gajovic-Eichelmann, and R. Hölzel, this volume (2002).
12. T. R. Strick, J. F. Allemand, D. Bensimon, A. Bensimon, and V. Croquette, Science **271**, 1835 (1996).
13. Bensimon, A. Simon, A. Chiffaudel, V. Croquette, F. Heslot, and D. Bensimon, Science **265**, 2096 (1994).
14. E. Braun, Y. Eichen, U. Sivan, et al., Nature **391**, 775 (1998).
15. R. M. Zimmermann and E. C. Cox, Nucleic Acids Research **22**, 492 (1994).
16. Wirde and U. Gelius, Langmuir **15**, 6370 (1999).
17. T. B. H. Vallant, U. Mayer, H. Hoffmann, T. Leitner, R. Resch, and J. G. Friedbacher, Physical Chemistry B **102**, 7190 (1998).
18. Csaki, R. Möller, and W. Fritzsche, Expert Review in Molecular Diagnostics **2**, 187 (2002).
19. G. Maubach, A. Csaki, D. Born, and W. Fritzsche, submitted (2002).
20. Csaki, R. Möller, J. Reichert, J. M. Köhler, and W. Fritzsche, in *Micro- and Nanostructures of Biological Systems*, edited by G. Bischoff and H.-J. Hein (Shaker Verlag, Aachen, 2001), p. 76.
21. J. Richter, R. Seidel, R. Kirsch, M. Mertig, W. Pompe, J. Plaschke, and H. K. Schackert, Advanced Materials **12**, 507 (2000).
22. W. E. Ford, O. Harnack, A. Yasuda, and J. M. Wessels, Advanced Materials **13**, 1793 (2001).
23. F. Monson and A. T. Woolley, this volume (2002).

DNA Templated Construction of Metallic Nanowires

Christopher F. Monson and Adam T. Woolley*
Department of Chemistry and Biochemistry
Brigham Young University
Provo, UT 84602-5700 USA
*Corresponding Author: atw@byu.edu

Abstract. Depositing silver, palladium or a silver-gold metal mixture onto surface immobilized DNA has made conductive nanowires in the 50–100 nm diameter range, a significant improvement over conventional optical lithography. Platinum has also been deposited onto DNA to form nanoscale metallic structure arrangements on surfaces. We have developed a method of elongating DNA for facile analysis on silicon or mica substrates. Silver and copper metal have both been deposited on these aligned surface DNA fragments to make nanorod structures that are ~2 nm in diameter, more than an order of magnitude smaller than earlier work.

I. INTRODUCTION

In the past few decades, the miniaturization of electronic components has allowed for faster and more tightly-packed circuits. Optical lithography has been the principal method used in the fabrication of integrated circuits, but it is now reaching its experimental limits. Current methods use soft UV radiation, and moving to even shorter UV wavelengths (which would enable smaller dimensions) is complicated by the ambient absorption of hard UV radiation by atmospheric gases. Thus, alternative methods of integrated circuit fabrication are attracting interest. One novel approach involves using DNA to position nanometer scale components and wires on surfaces. This technique can take advantage of the selectivity with which DNA base pairs bind to specifically locate nanostructures along the length of a DNA molecule on a surface. A critical aspect of this method lies in making segments of DNA conductive over long distances, as DNA is not a good intrinsic conductor [1]. Thus, attempts have been made to coat surface bound DNA with various metals, and herein we discuss recent advances in this field.

A number of different metals or combinations of metals have been deposited on DNA to make metallic nanorods: silver, palladium, platinum, copper, and a silver-gold composite. The nanostructures made using each of these metals will be discussed. Our group is studying the initial phases of metallic nanostructure formation on DNA templates. We have designed an approach for facile alignment of DNA fragments on a surface, and have made very thin (1-2 nm diameter) nanorods of silver and copper.

CP640, *DNA-Based Molecular Construction: International Workshop*, edited by W. Fritzsche
© 2002 American Institute of Physics 0-7354-0095-4/02/$19.00

II. STATE OF THE ART

Silver Nanowires

Silver nanowires were the earliest kind to be developed from a DNA template [2]. To make these nanostructures, Braun *et al.* attached thiol-modified oligonucleotides to gold electrodes separated by 12–16 μm. The oligonucleotides chosen were complementary to the 12-base sticky ends of λ DNA, which would then bridge the two electrodes when a solution containing λ DNA was flowed on the surface perpendicular to the electrodes. The λ DNA molecules specifically hybridized with the oligonucleotides at one electrode and were stretched in the flow direction until the other end of the molecule hybridized with the complementary oligonucleotides bound to the opposing electrode.

When the DNA bridge had been constructed, the surface was treated with a basic solution of silver nitrate. The positively charged silver ions associated with the negatively charged phosphate groups on DNA, and when hydroquinone was added under low-light conditions, the silver ions were reduced to silver metal and formed a granular silver coating on the DNA. When the reduction reaction was allowed to proceed until a linear feature was visible in an optical microscope, the nanowire became conductive. The nanowire conductivity was found to be ohmic in some areas of the current-voltage plot, and the nanowire behavior became ohmic throughout the measured range after a 50 V potential was applied to the electrodes. The silver nanorods made using this method could be made as small as a few nanometers in diameter, but did not become conductive until they were around 100 nm thick.

Palladium Nanowires

After this initial demonstration of DNA templated fabrication of silver nanowires, a technique for making palladium nanowires from a DNA template was published [3]. In this approach, DNA was immobilized on a glass surface by placing a droplet of a DNA solution on the substrate and allowing the droplet to evaporate. As the droplet evaporated, the DNA became aligned in the direction of the retreating liquid. The surface DNA was then treated with palladium acetate, which associated electrostatically with the DNA. Finally, palladium ions were reduced to palladium metal using a reducing solution consisting of sodium citrate, lactic acid, and dimethylamine borane [3].

In a subsequent publication, the conductivity of the palladium nanowires made using this method was reported [4]. Palladium nanowires were generated by the previously described procedure on top of a pair of comb-like gold electrodes on a glass surface. The nanowires were determined to exhibit ohmic behavior when their diameter was 50 nm or greater. Initially the resistance of the nanowires was greater than 5 kΩ, which was mostly due to contact resistance between the nanowires and the electrodes. Electron beam deposition of carbon lines over the wire-electrode interface reduced the contact resistance such that the resistance of individual nanowires fell below 1 kΩ [4]. These palladium nanowires were comprised of long chains of interconnected metallic grains, similar to the silver nanowires described earlier.

Platinum Nanostructures

Two different groups have created nanoscale platinum deposits on DNA. One approach [5] is an adaptation of the previously described method for palladium nanowire fabrication from DNA [3,4]. DNA was treated with a K_2PtCl_4 solution, and then a reducing agent, dimethylamine borane, was added. The DNA-cluster suspension was diluted with a HEPES–Mg buffer, placed on a freshly cleaved mica substrate for two minutes, and then rinsed with water. Deposition of small, ~5 nm diameter platinum clusters along much of the DNA was observed using both atomic force microscopy (AFM) and transmission electron microscopy (TEM). As a control, a K_2PtCl_4 solution was treated with the reducing agent and then DNA was added. This resulted in large, ~20 nm diameter platinum clusters that formed nonspecifically and deposited randomly onto the entire surface. In another control experiment, K_2PtCl_4 solution, reducing agent, and DNA were all added at the same time. This resulted in an intermediate behavior, with some platinum nanoparticles associating along the DNA and some forming at random locations. These observations indicated that the platinum associated with the DNA and had been reduced while still bound to the DNA. The resulting platinum nanoparticles then acted as catalytic sites for the reduction of the remaining platinum in solution, resulting in the formation of aligned platinum nanostructures [5].

In an alternative approach to making platinum nanostructures, Ford *et al.* exposed DNA to two different forms of platinum, $[Pt(terpy)Cl]Cl \cdot 2H_2O$ and *cis*-$Pt(NH_3)_2Cl_2$ [6]. The unassociated platinum was removed from the solution via ion exchange, and borohydride was added to reduce the platinum, resulting in the formation of platinum metal aggregates along the DNA. The deposition of DNA onto a substrate was done both before and after platination. In either case, silicon was reacted in an O_2 plasma and then mounted onto a spin-coater. Several drops of DNA or Pt/DNA solution were placed on the wafer, and after two minutes the sample was spun, being rinsed with water while still spinning to remove any salt. The DNA/platinum nanostructures were then imaged with AFM and TEM, where aggregates with diameters from one to a few nanometers were observed. The coverage of platinum along the DNA was not complete, although it was suggested that the platinum clusters could serve as catalytic sites for the deposition of gold or other metal nanoparticles. This second metal deposition could complete the coating and form conductive nanowires.

Silver-Gold Composite Nanowires

Recently, DNA templated construction of silver-gold composite nanowires has been reported in a two-step procedure [7]. First, surface-bound DNA was treated with Ag^+, leading to the formation of silver aggregates along the DNA. Next, the DNA was treated with $KAuCl_4$, KSCN, and hydroquinone, which reduced the gold ions to gold metal; the silver aggregates along the DNA catalytically promoted the selective deposition of gold along the surface of the DNA molecules. These silver-gold composite nanowires displayed ohmic behavior, with a resistivity about an order of magnitude above that of polycrystalline gold (1.5×10^{-7} ohm·m for the nanowire, 2.2×10^{-8} ohm·m for polycrystalline gold).

In the same paper [7] the DNA-templated creation of conducting nanowires with an insulating gap was also reported. To make these nanostructures, single-stranded DNA was treated with RecA protein, which polymerized on the DNA, acting as a protecting group. The RecA-protected single-stranded DNA was bound to an aldehyde-derivatized double-stranded DNA molecule at a homologous sequence, and the composite DNA-protein complex was treated with silver and gold as before. AFM and scanning electron microscopy revealed ~50 nm diameter nanowire formation along the DNA length, except for the RecA-protected section.

III. EARLY STAGES OF NANOWIRE FORMATION

DNA Alignment

While the alignment of double-stranded DNA on surfaces is fairly straightforward (for example, several different methods were used to make the previously discussed DNA templated nanostructures), the alignment of well-extended single-stranded DNA presents greater difficulties. Our group has developed a method that works well for aligning and extending both single- and double-stranded DNA on surfaces [8]. This approach takes advantage of the surface forces from a retreating liquid meniscus as a droplet of DNA solution is moved across a surface. Silicon or mica surfaces are treated with poly-L-lysine (0.2–10 ppm in water), which gives the surface a positive charge to which negatively charged DNA will be electrostatically bound. A 1 μL droplet of DNA solution (0.5–5 ng/μL) is then translated across the surface using a three axis translation stage. The resulting surface forces align well-extended single- and double-stranded DNA fragments on the surface.

This same approach has also been used to generate more complex aligned DNA surface patterns. Single-stranded DNA has been aligned orthogonal to double-stranded DNA on a surface, indicating the robustness of the surface attachment and its stability in multistep surface preparations [8]. The ability to readily generate well attached, aligned and extended DNA fragments on surfaces has proved invaluable in our subsequent work involving the deposition of metal from solution onto aligned surface DNA molecules.

Silver Nanorods

Using DNA aligned on a silicon surface by our previously described method, we have investigated the early stages of metal nanowire formation on DNA templates. DNA on a surface was treated with 0.11 M $AgNO_3$ solution and then briefly (~10 min) exposed to light to photoreduce the silver. The Ag^+ solution was removed and the surface was rinsed and imaged. These Ag^+ treatments could be repeated for further silver deposition along the DNA.

Surfaces prepared in this fashion were imaged by AFM, and the height of the DNA nanostructures relative to the surface was measured before and after each silver treatment. The average height of the DNA increased by 0.25 nm after the first silver treatment, and a further 0.22 nm after a second treatment (see Table 1). A modification

TABLE 1. Average DNA height after various treatments

	DNA[a]	1 x KNO_3	2 x KNO_3	First Method 1 x $AgNO_3$	First Method 2 x $AgNO_3$	Black Box Method 1 x $AgNO_3$	Black Box Method 2 x $AgNO_3$	Black Box Method 3 x $AgNO_3$	Black Box Method 4 x $AgNO_3$	1 x CuCl
Height (nm)	1.196	1.288	1.339	1.444	1.662	1.508	1.611	1.701	1.735	1.350
St. dev. (nm)	0.322	0.359	0.381	0.466	0.677	0.411	0.431	0.389	0.454	0.483

[a]Untreated λ double-stranded DNA

In all cases, at least 100 measurements were taken. For some cases, significantly more measurements were taken (for the untreated DNA and single $AgNO_3$ treatment using the first procedure developed, 359 and 244 measurements were taken, respectively).

of the above method was developed, in which the silver was allowed to associate along the DNA for 5–10 min before being exposed to light (the "black box" method). This approach resulted in metal deposition along the DNA that was more specific (see Figure 1), and substantial (the first silver treatment using the black box method caused a height increase of 0.31 nm compared to 0.25 nm). Because of the increased specificity of the silver deposition, as many as four treatments were possible using the black box method, before nonspecific deposition to begin to obscure the DNA (see Table 1). As a control experiment, DNA on a surface was subjected to multiple treatments with KNO_3 in the same manner as before. This caused only very slight increases in height, confirming that the silver was indeed responsible for the observed height increases in the previous results.

DNA that had been treated with silver began to have a granular appearance, especially when multiple treatments had been performed. This result is consistent with other work in metallization of surface DNA that has been discussed previously. In our work, the silver aggregates are extremely small, with maximum diameters of ~ 2 nm and typical diameters less than 1 nm (see Figure 2). In some regions of the DNA, it appeared that these silver nanoaggregates almost completely covered the DNA, especially after multiple treatments.

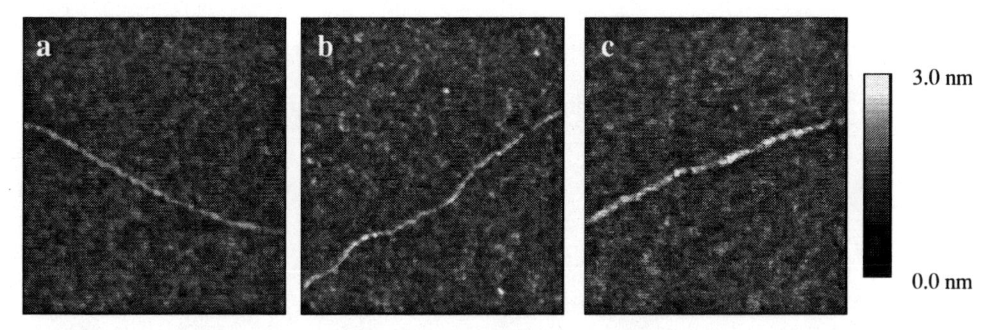

FIGURE 1. Tapping mode AFM images of λ double-stranded DNA. (a) DNA elongated and attached to the surface, but otherwise untreated, shown for comparison. (b) DNA treated with $AgNO_3$ once in full light. The deposition is not entirely specific, with some deposition occurring on the silicon surface. (c) DNA treated with $AgNO_3$ once using the "black box" method. Note the increased feature height and lack of nonspecific deposition compared to (b). All images are 500 x 500 nm, and the height scale is shown at the far right.

FIGURE 2. Tapping mode AFM images of λ double-stranded DNA on silicon. (a) Untreated DNA has a relatively uniform height. (b) DNA that has been treated with $AgNO_3$ three times has an increased height, as well as a much more granular appearance, indicating areas of metal deposition. In both images, the scale bar indicates 100 nm.

Copper Nanorods

We have also been evaluating DNA templated synthesis of copper nanorods. Our initial efforts focused on the disproportionation of copper (I) to form copper (II) and $Cu_{(s)}$. DNA aligned using previously developed procedures [8] was treated with a 40 μL droplet of saturated aqueous CuCl solution, sometimes following a treatment with silver nitrate as previously discussed. Copper (I) ions disproportionated to form copper metal and copper (II) ions; some copper metal was deposited along the DNA, while the copper (II) remained in solution. The deposition of copper metal along the DNA resulted in an increase in the average height of the DNA of 0.15 nm (see Table 1), corresponding to the deposition of slightly more than one-half of a monolayer of copper atoms on the DNA template.

Deposition of metallic copper by disproportionation of copper (I) occurred not only on the template DNA, but also on other random surface locations. Because of this, only one copper deposition cycle was possible, because further copper treatments also increased the height of the non-specific copper deposits, such that visualization of the DNA templated Cu nanorods became difficult by AFM. Thus, alternate methods will need to be explored to deposit sufficient copper on surface DNA fragments to form conducting nanowires.

IV. CONCLUSIONS

Using DNA templates, conductive nanowires have successfully been made using silver, palladium, and a silver-gold mixture. These nanowires are still 50–100 nm wide, but this nonetheless represents an improvement over the current optical lithographic methods. Platinum has been deposited on DNA in such a manner that conductive nanowires could likely be generated. We have developed a robust method for localization and alignment of DNA on surfaces. Using these template nucleic acids we are studying the initial stages of metallic nanowire formation. We have characterized DNA templated nanorods of silver and copper having cross-sections less than 2 nm.

Future conductivity studies of these nanostructures will offer insights into the lower dimensional limits for DNA templated nanowire synthesis.

Future Directions

New methods of making nanowires on DNA, both from metals described herein and new materials, should be developed to allow thinner conductive nanowires to be created. For example, conductive carbon nanotubes could be specifically localized on DNA instead of metal; alternatively, multiple depositions of metal nanoparticles onto surface DNA could be explored. Moreover, methods of combining DNA with nanodevices should be tested in an effort to make biotemplated circuits. Keren *et al.* offer a promising first step in this direction with the construction of DNA templated nanowires containing an insulating gap [7]. From this starting point, it should be feasible to generate simple DNA templated nanocircuits, and the creation of more sophisticated nanoscale integrated circuits becomes a distinct possibility.

ACKNOWLEDGMENTS

Supported in part by startup funds from Brigham Young University. We also acknowledge the donors of The Petroleum Research Fund, administered by ACS, for partial support of this research. C. Monson was supported by an Undergraduate Research Award of the College of Physical and Mathematical Sciences, Brigham Young University.

REFERENCES

1. Storm, A. J., van Noort, J., de Vries, S., and Dekker, C., *Appl. Phys. Lett.* **79**, 3881–3883 (2001).
2. Braun, E., Eichen, Y., Sivan, U., and Ben-Yoseph, G., *Nature* **391**, 775–778 (1998).
3. Richter, J., Seidel, R., Kirsch, R., Mertig, M., Pompe, W., Plaschke, J., and Schackert, H. K., *Adv. Mater.* **12**, 507–510 (2000).
4. Richter, J., Mertig, M., Pompe, W., Mönch, I., and Schackert, H. K., *Appl. Phys. Lett.* **78**, 536–538 (2001).
5. Seidel, R., Mertig, M., and Pompe, W., *Surf. Interface Anal.* **33**, 151–154 (2002).
6. Ford, W. E., Harnack, O., Yasuda, A., and Wessels, J. M., *Adv. Mater.* **13**, 1793–1797 (2001).
7. Keren, K., Krueger, M., Gilad, R., Ben-Yoseph, G., Sivan, U., and Braun, E., *Science* **297**, 72–75 (2002).
8. Woolley, A. T., and Kelly, R. T., *Nano Lett.* **1**, 345–348 (2001).

DNA SUPERSTRUCTURES

Assembly Of G-Quartet Based DNA Superstructures (G-Wires)

Anett Sondermann, Claudia Holste, Robert Möller, Wolfgang Fritzsche

Institute for Physical High Technology Jena, Germany, Biotechnical Microsystems Department
fritzsche@ipht-jena.de

Abstract. G-wires are DNA superstructures based on the intermolecular interactions of four Guanine bases. They allow the fabrication of structures reaching the micrometer scale using only short DNA oligonucleotides, what makes them potentially interesting for molecular nanotechnology. We investigated the assembly of G-wires by SFM, using different sequences described in the literature. The assembled structures were adsorbed on mica and imaged by SFM. The influence of time and temperature of the growth was investigated, and the topology of the assemblies was studied.

INTRODUCTION

Self-organization of short molecular units into larger structures is a basic method in molecular nanotechnology. DNA units are especially promising due to their large variety in sequence and the great number of established techniques for manipulation and characterization. The hybridization of two DNA molecules with complementary sequences is a key contribution of DNA technology to the field of molecular nanotechnology. By defined variations in the sequences, a variety of different coupling pairs can be created, which bind in a parallel manner under similar physicochemical parameters – in contrast to combinations of other known binding systems as e.g. biotin- (strept) avidin, thiol-gold or his-tag. It allows in principle the fabrication of structures in the micrometer range of known sequences starting from easily accessible short (<100 bases) molecules as an important step for the interfacing of molecular constructs with microstructured surfaces as a technological environment. However, many different molecules, several hybridization steps as well as a sophisticated protocol are needed for the successful implementation of this approach. Beside these complications, the resulting double-stranded DNA-molecule exhibits probably hardly any electrical conductivity. So other schemes are needed to achieve molecular wires in the micrometer range by DNA technology.

One promising approach is based on G wires, a DNA superstructure of guanine-rich oligonucleotides. They are formed by intermolecular interaction of the guanines resulting in quartets of guanines (G-quartets) stabilized by hydrogen bonds. G-quartets are probably used for a stabilization of the overlapping single-stranded ends of chromosomes, which exhibit typically a G-rich sequence. This effect is realized by back folding of the strand resulting in intramolecular G-quartets. Using this assembly approach (but intermolecular quartets), superstructures made from short

CP640, *DNA-Based Molecular Construction: International Workshop,* edited by W. Fritzsche
© 2002 American Institute of Physics 0-7354-0095-4/02/$19.00

oligonucleotides containing guanines were realized and characterized using spectroscopic and microscopic methods [1-3]. Typically thread-like structures with lengths in the upper nanometer and even micrometer range are observed. Here we report our results of a microscopic characterization of G-wires after growth in solution and subsequent adsorption onto mica substrates.

MATERIALS AND METHODS

G-Wire Assembly

A G-wire sequence (G4, [1]) and two derivates were studied: G2 (G2T2G2), G4 (G4T2G4), and G6 (G6T2G6). 125 μM DNA was incubated in buffer A containing 50 mM NaCl, 10 mM $MgCl_2$ and 50 mM Tris-HCl at pH 7.5 for up to 10 days and at a constant temperature in the range 21-50° C.

Microscopy

Scanning force microscopy (SFM) was applied for a visualization of the grown assemblies. Therefore, samples from the buffer A were diluted 1:100 in a buffer B containing 1 mM $MgCl_2$ and10 mM Tris-HCl at pH 7.5. From the resulting solution, a 3 μl droplet was deposited on a freshly cleaved mica substrate. After a 5 min. incubation, the substrate was washed with 1 ml distilled water prior to drying in a stream of nitrogen. For imaging, a scanning force microscope NanoScope III (Digital Instruments, Santa Barbara, CA) with either a Dimension 3100 or a Multimode head was used in the tapping mode.

RESULTS AND DISCUSSIONS

The defined positioning of DNA molecules on solid substrates will be an essential element for the creation of DNA-based devices and materials, as envisioned by DNA nanotechnology. Although short molecules can be bind on one end following techniques developed for Self Assembly Monolayer (SAM) studies or DNA chips, the number of molecules is usually not defined in these experiments. The use of longer DNA in the micrometer range limits the concentration due to steric hindrance, so that individual molecules can be observed. While techniques like optical tweezers use these molecules in standard procedures, the surface binding is rather difficult as the small number of publications in this field implies. One possible reason is the high flexibility of DNA, especially under the influence of a moving air-water interface during the drying procedure. Another related point is the missing adhesion of the DNA strand to the substrate, which could stabilize against the high forces exerted by the moving interface.

DNA superstructures created by self-assembly of molecular units would combine the advantages of both approaches. Such structures could be designed to exhibit a higher stiffness and therefore an increased stability, connected with the ease of immobilization as well as manipulation. Seeman et al. pioneered this approach demonstrating a variety of structures made out of DNA units [4]. Beside these structures made of several different units, DNA superstructures based on only one sequence of oligonucleotide were reported. These structures are preferably four-stranded and contain G-quartets as connecting motif [1].

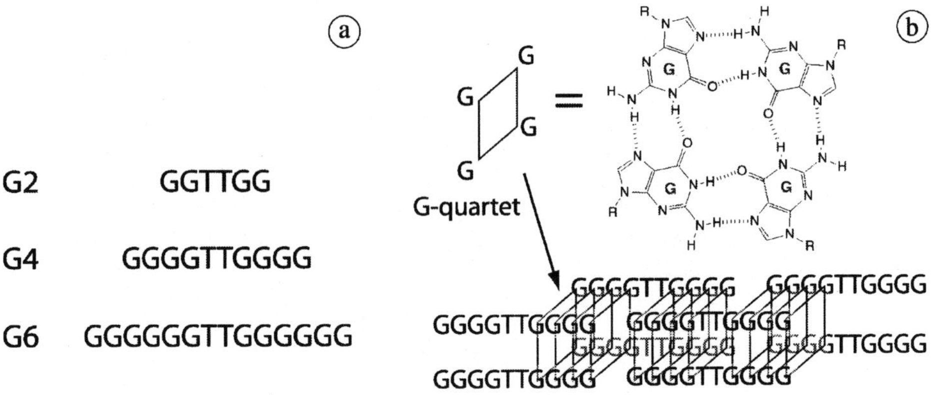

(a)

G2 GGTTGG

G4 GGGGTTGGGG

G6 GGGGGGTTGGGGGG

(b)

G-quartet

FIGURE 1. a) Sequences investigated in this study. b) Scheme of proposed G-wire structure [1]

DNA Sequence

The DNA unit for a G-quartet based self-assembly should contain guanine bases. Three sequences were chosen from the literature, which all exhibited two thymine bases flanked on each side by 2-6 guanines (Fig. 1a). A standard protocol was used to induce G-wire growth. After certain times (3, 6 and 10 days, respectively) samples from the solution were prepared for SFM by surface adsorption onto freshly cleaved mica prior to imaging. The visualized samples are shown in Fig. 2.

The first sequence (G2) shows no significant amount of superstructures in the observed timeframe of 10 days (Fig. 2a). Only occasionally the smooth mica surface reveals some structures, which can be attributed to artifacts probably based on dried salt residues from he buffer solution.

The picture is quite changed in the case of the second sequence (G4) shown in Fig. 2b. Already after 3 days, the surface is covered by thread-like structures in the lower nanometer length range, occasionally reaching up to 150 nm (top). With longer incubation time, the share of smaller structures seems to decrease (center). Finally, after 10 days structures of up to 400 nm length become visible (bottom).

Images of adsorbed samples from the third sequence (G6) revealed large deposition of material (Fig. 2c), which exhibit occasionally a homogenous thickness but with visible holes (large structure in the upper image). A possible cause could be salt residues from the used buffer. However, the same buffer was used for all three sequences, so that salt alone can be excluded due to the missing features in Fig. 2a and b. So a DNA-related process seems to be responsible. Sheet-like structures were described in the literature for G-wire preparations, but looked quite different to our observations [5]. We assume that the longer guanine flanks of G6 are able to more complex interactions compared to G4, and that this effect is responsible for the observed phenomenon. Further detailed experiments will aim at the influence of extended G-flanks onto the observed structures.

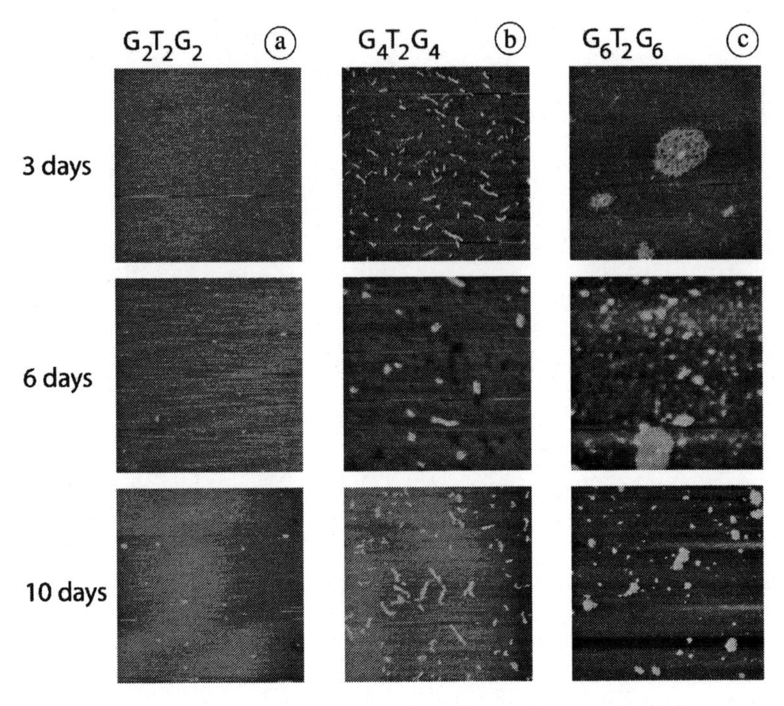

FIGURE 2. Influence of the DNA sequence onto the self-assembly properties. Sequences with two central thymines and flanked by 2 (a), 4 (b) or 6 (c) guanines were incubated for 3-10 days and visualized by SFM after adsorption onto mica (scan size 2x2 μm).

Incubation Temperature

The influence of the incubation temperature on the G-wire growth from the promising sequence G4 was studied. 21° C and 50° C were tested, beside 37° C as suggested by the literature.

The sample incubated at 21°C showed no DNA assemblies at all (Fig. 3a). The experiment with 37° C confirmed the results from the experiments with different sequences described in Fig.2b: The appearance of thread-like structures on the substrate, with a density increase over time (Fig. 3b, c). The sample incubated at 50° C revealed already after one day some contrast in the SFM, but only after three days thread-like structures became apparent. Compared to the structures observed at 37° C, the G-wires are much longer (more than 600 nm), and they are more often branched. The thickness of the G-wires is not constant especially in the image after 3 days, an observation which could be an artifact caused by adsorption effects or technical problems related to humidity and tip shape. If this inhomogeneous diameter is not a measurement artifact, it could point to a broadening of the typical self-assembly structure given in Fig. 1b: The drawing describes an assembly with one central axis through the G-quartets, and a growth only in the directions of this axis. A broadening would occur in the case of sideward growth, resulting in new G-quartets beside the given axis. Such a mechanism could also explain the observation of branches by partial overlap on two sides of the original axis.

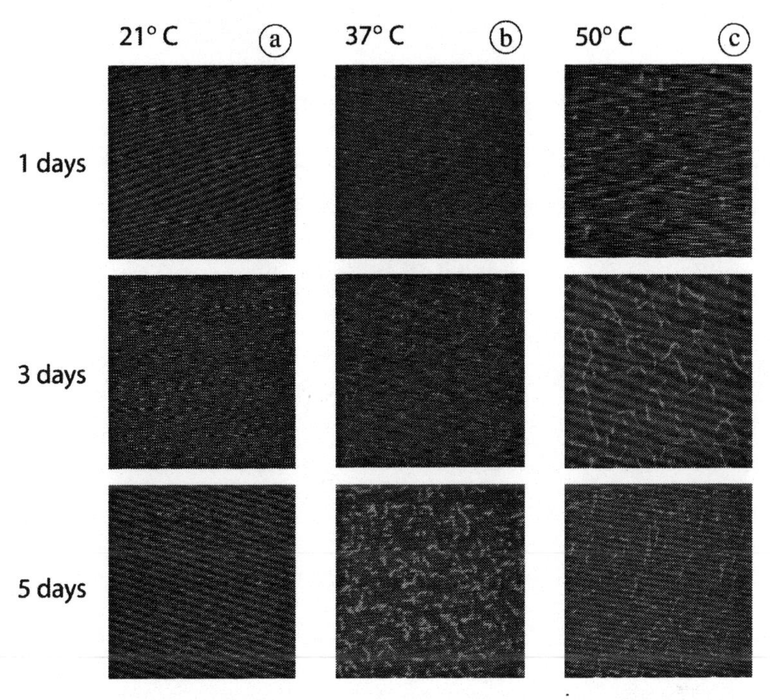

FIGURE 3. Temperature dependency of the assembly of G-wires from G4. The DNA was incubated at 21°C (a), 37° C (b) or 50° C (c) for 1-5 days. The grown assemblages were adsorbed onto mica and imaged by SFM (scan size 2x2 μm).

Alignment

Molecular nanotechnology needs molecular units but also tools accessing and manipulating these parts. Self-assembly strategies are a hopeful alternative to slow serial manipulation methods like optical tweezers or scanning probe techniques. So any topological control over the adsorption of molecular assemblies is of interest as potentially interesting construction scheme. In this regard the observation of G-wire orientation along the crystal axes of mica was promising [5]. Figure 4a shows a zoom of adsorbed G-wires, where the preferred adsorption along the three axes (separated by 120° each) becomes apparent. One has to point out that this orientation occurs after the growth of the G-wires, the grown assemblages are apparently registering in the moment of adsorption from the liquid phase. The axes of mica are not only visualized by G-wire decoration; Fig. 4b shows similar structures in a negative contrast after a mica surface was treated with a buffer solution without DNA. This image points to other sources for the observation of oriented structures on the surface beside DNA, emphasizing the needed careful consideration of results in this area.

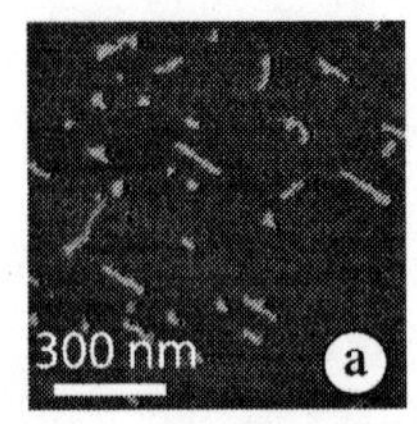

FIGURE 4. Orientation effects on mica surfaces visualized by SFM. a) G-wires adsorb in three distinct orientations on mica substrates, which are separated by 120° each. b) Observation of similar structures but in a negative contrast after a mica surface was incubated with buffer B (without DNA).

Conclusions

G-wires are an interesting DNA superstructure that can be easily assembled and adsorbed onto surfaces. If the proposed, potentially interesting electronic properties [5] are confirmed, G-wires could become another useful tool in the growing set of techniques for a molecular nanotechnology and –electronics.

ACKNOWLEDGMENTS

We thank J. Vesenka, E. Henderson and T. Marsh for introduction into the field and providing information and unpublished material, S. Diekmann and M. Fändrich for access to the Multimode SFM at the IMB, M. Kittler and A. Csaki for assistance with SFM measurements. We acknowledge J. M. Köhler for his initial and essential contributions to the establishment of molecular nanotechnology at the IPHT.
This work was supported by the DFG (FR 1348/3-4) and the Volkswagen Foundation (Priority Area: Physics, Chemistry and Biology with Single Molecules).

REFERENCES

1. T. C. Marsh, J. Vesenka, and E. Henderson, Nucleic Acids Research **23**, 696 (1995).
2. F. Sha, R. Mu, D. Henderson, and F. M. Chen, Biophys J **77**, 410 (1999).
3. T. Muir, E. Morales, J. Root, I. Kumar, B. Garcia, C. Vellandi, D. Jenigian, T. Marsh, E. Henderson, and J. Vesenka, Journal of Vacuum Science and Technology A **16**, 1172 (1998).
4. N. C. Seeman, Annu Rev Biophys Biomol Struct **27**, 225 (1998).
5. J. Vesenka, E. Henderson, and T. Marsh, this volume (2002).

Construction and Examination of “G-wire” DNA

James Vesenka[a], Eric Henderson[b] and Thomas Marsh[c]

[a]University of New England
11 Hills Beach Road
Biddeford, ME 04005
Email: jvesenka@une.edu
[b]BioForce Nanosciences, Inc.
2901 South Loop Drive, Suite 3400
Ames, IA 50010 USA
[c]University of St. Thomas
2115 Summit Avenue, St. Paul, MN 55105

Abstract. The self-assembly process of G-wire DNA was investigated through scanning probe microscopy. Growth kinetics studies indicated the self-assembly process is diffusion limited and provides Poisson-like distribution of G-wire lengths upon reaching equilibrium. This evidence suggests that self-assembly is driven by thermodynamic processes. The average lengths of these molecules are around 100 nm long after 24 hours of growth. However, longer G-wire DNA molecules (many micrometers) are found both in flexible and crystalline forms. The latter structures are extremely interesting candidates for molecular nanowires.

INTRODUCTION

G-DNA is a polymorphic family of four-stranded structures containing guanine tetrad motifs (G-quartets) [1,2]. Guanine rich oligonucleotides that are self-complimentary, as found in many telomeric (chromosome ends) repeat sequences, form G-DNA in the presence of monovalent and/or divalent metal cations. The atomic force microscope (AFM) and low current scanning tunneling microscope (LCSTM), high resolution, near-field, three dimensional imaging devices, were used to explore the growth of linear G-4 polymers. These “G-wires” [3] are speculated to form by the self-assembly of the telomeric oligonucleotide sequence d(GGGGTTGGGG) also called d“Tet1.5” monomers (Fig. 1). One of the exciting early observations was the topographic difference between G-wire and duplex DNA (Fig. 2). Whereas duplex DNA tends to collapse on the surface of mica, G-wire DNA appears to hold its shape [4]. Hydration low current scanning tunneling microscopy [5] has provided evidence of semi-conductivity [4]. This stability may be the result the base stacking of G-quartets and caged monovalent cations [3,6] (Fig. 3). It is this stability, uniformity and long lengths that make G-wire DNA templates for molecular wiring [7,8]. Little is known about the optimal conditions needed to construct long wires. This study attempts to develop baseline conditions for the growth of G-wire DNA, and estimates the equilibrium constant for multimers, in an effort better control the growth of “superstructures” (many micrometers long).

CP640, *DNA-Based Molecular Construction: International Workshop,* edited by W. Fritzsche
© 2002 American Institute of Physics 0-7354-0095-4/02/$19.00

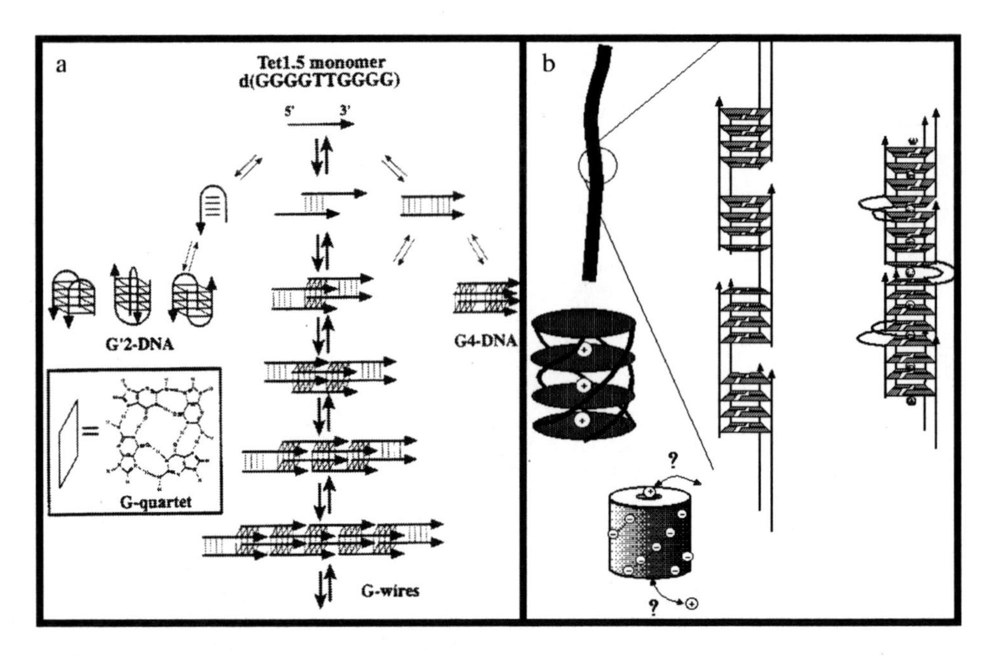

FIGURE 1. The stability of the G-wire DNA is better understood through examination of the hypothesized growth mechanism. The Tet1.5 monomer can form a dimer pair with either a "closed", "looped", or "staggered" conformations as shown in Fig. 1a. In either of the closed or looped conformations no more growth of the G-wires can occur. In the staggered conformation another dimer can attach to the G-wire ladder creating a succession of "sticky ends", enabling multimers to assemble. The process is driven thermodynamically (see text). Monomeric cation species such as potassium or sodium are thought to help stabilize the G-wires down the base-stacked core of the structure as seen in Fig. 1b [3,7]. The thymine groups may act as flexible links that can "bunch up" in solution or after adsorption onto a substrate.

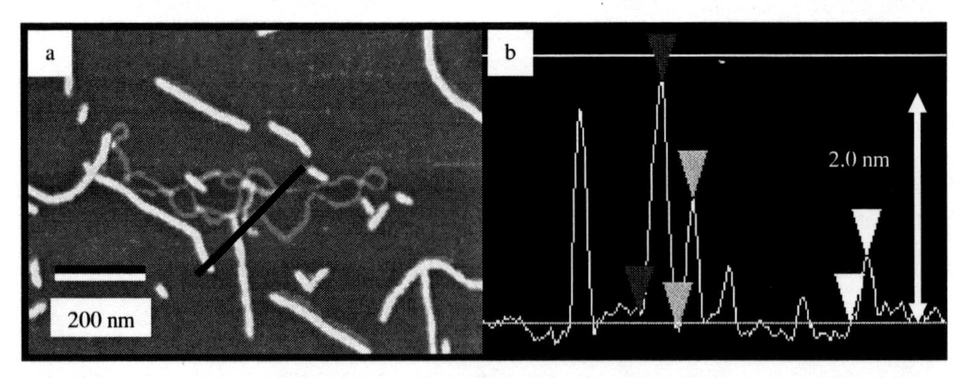

FIGURE 2. Topographic comparison between duplex and quadruplex "G-wire" DNA co-adsorbed onto the same substrate (Fig. 2a). The duplex DNA collapses on the surface of mica to a height of 0.5 nm above the surface as seen in the cross section of Fig. 2b. Even the supercoiled segments of the double stranded DNA measure only about 1.0 nm above the mica substrate. This is about half the diameter expected from Watson-Crick duplex DNA in solution. However, the quadruplex DNA is uniformly about 2.2 nm in diameter, very close to the NMR and x-ray spectra of G-quartet DNA (2.4 nm in diameter.) Vertical height range is 10 nm from black to white.

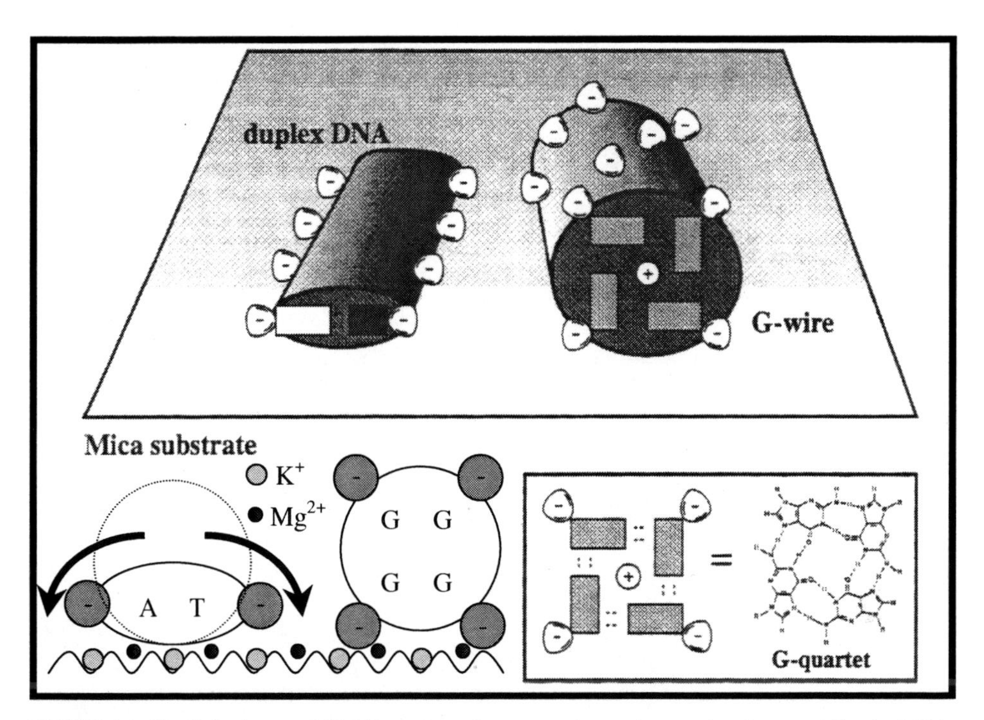

FIGURE 3. The "pinning model" [4] suggests the greater internal attractive forces of G-wire DNA, comprised of guanine-quartet building blocks, four in a row, enable it to retain its solution state structure when exposed to tether cations, here shown as magnesium. However, the stronger tethering force exerted on the unsupported phosphate backbone of duplex DNA, shown here with an example A-T base pair, pins the duplex DNA flat to the mica substrate.

MATERIALS AND METHODS

Quadruplex G-wire DNA was prepared by according to the procedure outlined by Marsh et al. [3]. Temperature control of the $G_4T_2G_4$ monomers (Tet1.5) was maintained with a PCR Thermocylcer (Thermo Hybaid, U.K.). The solution is raised to 95°C for ten minutes to promote the melting of fortuitous G-4 structures, i.e. to ensure a cocktail consisting of monomers. Samples of the concentrated G-wire DNA (monomer concentration 1.0 mM) were diluted to a factor of 10 to 100 in a buffer consisting of 10 mM Tris (pH 7.6), and 1 mM MCl_x, where M = sodium, potassium, magnesium or zinc. These samples were allowed to sit on Parafilm for ten minutes at room temperature and adsorbed onto freshly cleaved muscovite mica. The samples were then incubated on the mica between 0 to 10 minutes rinsed with 1 ml deionized water, dried in a stream of dry nitrogen and allowed to stabilize in a 37°C oven before imaging. Samples were imaged with 125 µm x 20 µm silicon nitride probes using a Nanoscope E controller (Digital Instruments, Santa Barbara CA) in contact mode in dry air. Fresh samples were imaged under ambient humidity using a Nanoscope IIIa controller in TappingMode™ and 75 µm long Tapping Tips. Freshly prepared G-wire

DNA was also imaged with a PicoSPM (Molecular Imaging, Tempe AZ) low current STM in humid air.

RESULTS AND DISCUSSION

Fig. 4 is an example of a concentrated G-wire network that was created by depositing a sample containing a concentrated 24 hour-old G-wire solution onto freshly cleaved mica. The sample is rinsed and immediately imaged in TappingMode™ revealing an oriented network of G-wire strands over the surface of mica in Fig. 4a. The orientation affect is due to G-wire alignment with potassium vacancy sites on the surface of freshly cleaved mica (unpublished data). The density of the G-wire DNA appears to depend upon local variations of the mica surface. For example, imaging a region of the mica surface a millimeter away can provide results similar to Fig. 4c in which the G-wires are clearly separated. In Fig. 4b the same sample from Fig. 4a had been dried for 24 hours. Note that the G-wire DNA appears much narrower because of the dehydration of the hydration layer over the surface. The hydration layer is absolutely essential for imaging of the molecules via low current scanning tunneling microscopy (LCSTM) seen in Fig. 4c. Heim et al. [5] have called this form of LCSTM "hydration layer scanning tunneling microscopy" to account for the electrical conductivity over the sample's surface.

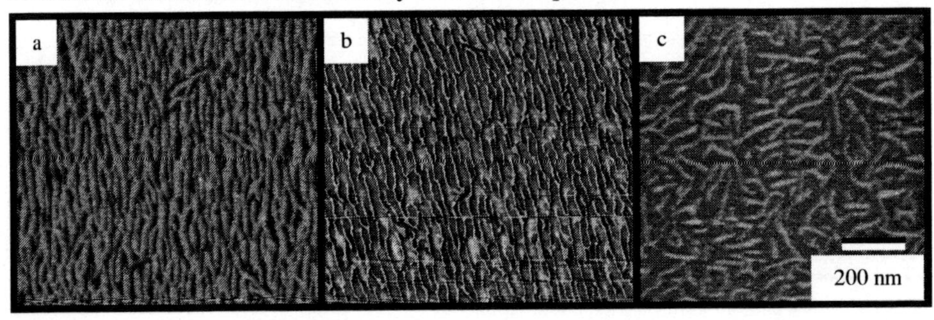

FIGURE 4. G-wires freshly adsorbed onto mica imaged via tapping mode (Fig. 4a) and the same sample imaged by the same tip 24 hours later after drying in an oven at 37°C (Fig. 4b). Note the preferential orientation is NOT a sample preparation artifact, e.g. due to rinsing. G-wires appear to align with the underlying potassium vacancy sites of the freshly cleaved mica substrate. Note the broadening of the G-wire structure due to hydrated residual buffer salts. After drying in an oven the G-wires appear much narrower and the buffer salts appear to distribute themselves in between the DNA strands. It is the freshly made, hydrated form of the G-wires, that are essential for low current scanning tunneling microscopy (LCSTM) imaging seen in Fig. 4c. This LCSTM image of G-wires freshly adsorbed on mica is recorded at 1pA tunneling current, -7V bias and 80% relative humidity in a sealed imaging chamber. Successful images of G-wires on mica occur most frequently on freshly prepared samples, high G-wire densities (but NOT G-wire "networks" as seen in Figs. 4a and b), and high concentrations of residual buffer cations. Note in this image sample there are pathways of tunneling current that does not require travel over the G-wires. These pathways are important for providing electron conduction from bias voltage source. Vertical height range is 10 nm from black to white.

The proposed growth process is described in Fig. 5. The Tet1.5 monomers are grown in a medium containing monovalent cations (either potassium or sodium) and divalent cations (typically magnesium). The monovalent cations appear to stabilize the G-wire internally by residing in the hole created by the four G-quartet envelope, also seen in Fig. 1b [6]. The divalent cations appear to stabilize the structure externally by residing between phosphate groups along the four-stranded backbone. The monomers combine to form the ladder structure L_1, or the staggered structure L_1', with associated equilibrium constants K_1 or K_1'. The rate-determining step in the growth process of the G-wires is the generation of L_2, an essential structure required for the growth of extensible G-quartet multimers. After the construction of L_2 the growth of longer wires is possible. Initially the concentration increase of L_2 is driven thermodynamically:

$$\Delta G = \Delta G^o + RT\ln(Q) \tag{1}$$

where the reaction quotient "Q" is defined by

$$Q = a_{L2}/a_{L1}^2 \approx [L_2]/[L_1]^2 \tag{2}$$

Here "a_{L2}" is the activity of the dimer and "a_{L1}" is the activity of the monomer. The ratio of the activities is approximated by the ratio of their respective concentrations.

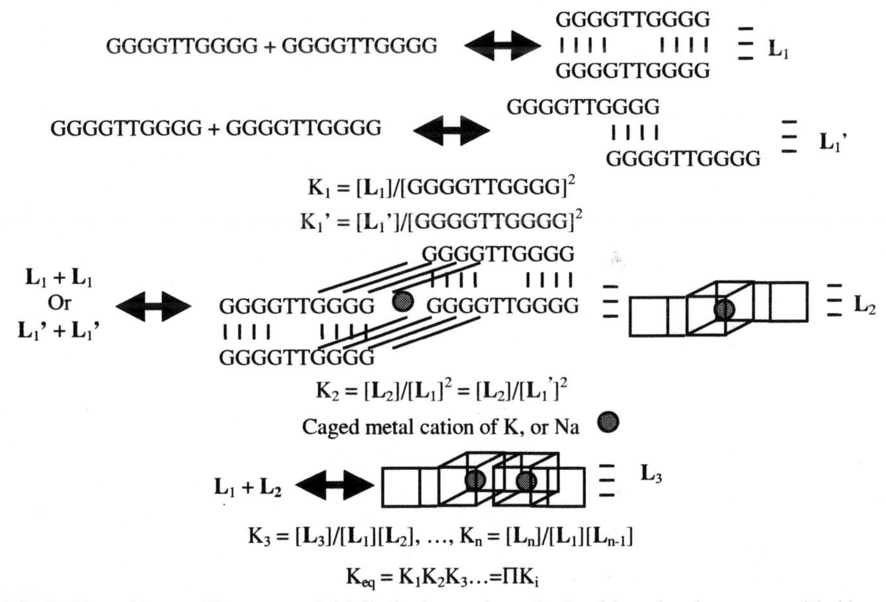

FIGURE 5. The self-assembly process initially is thermodynamically driven by the excess of ladder monomers L_1 and L_1'. As the concentration of dimers (L_2) reaches equilibrium with the monomers, the process of growing larger G-wire structures is driven by the excess numbers of dimers and monomers compared to the smaller concentrations of multimers (L_n). Characterization of the equilibrium constant for different length molecules involves the product of the equilibrium constant of the different length multimers. Complicating the growth process even more is the caged metal ions, sodium and or potassium. Incorporation of potassium leads to a more uniform spread of smaller G-wire DNA, whereas incorporation of sodium leads to a greater distribution in G-wire lengths.

The free energy initially drives formation of the dimers because the concentration of L_2, and thus Q, is zero. Consequently ΔG is large and negative (spontaneous assembly). The system rapidly seeks to reach equilibrium between dimers and monomers. Eventually the concentration of L_2 increases sufficiently that the system approaches equilibrium and the growth of longer wires is thermodynamically driven by the incorporation of dimers into the much lower concentrations of multimers, Fig.5.

The formation of dimers is always spontaneous because of the competition between polymerization into multimers with decomposition into L_1 or L_1'. The total self-assembly equilibrium constant, K_{eq}, can be expressed as a product of all the individual equilibrium constants for a given length G-wire multimer $\mathbf{L_n}$:

$$K_{eq} = \Pi K_i = [L_n]/[G_4T_2G_4]^2[L_1]^{n-1} \approx [L_n]/[G_4T_2G_4]^{n+1} \quad (3)$$

where the last approximation is the result of assuming the concentrations of Tet1.5 and L_1 are about the same.

If the model is accurate we would expect to see a Poisson-like distribution of G-wire DNA lengths, with a greater number of smaller length G-wires compared to longer structures. Furthermore, if the process is diffusion limited by the concentration of monomers interacting with multimers we would expect a comparison of length versus time to behave in a square root of time dependence. Both of these features can be seen in Fig. 6. In the time study presented in Fig. 6 the mean length <L> of the empirically determined growth rate is:

$$\langle L\rangle(\text{nm})=28\text{nm}+[(11\text{nm})/\sqrt{\text{day}}]\sqrt{t} \quad (4)$$

where "t" is time in days. The appearance of short wires happens almost immediately at the concentrations undertaken in this study, i.e. 1.0 mM Tet 1.5 monomer (Fig. 6a). Attempts to resolve mean lengths at smaller time scales (seconds) always yield short length G-wire DNA. At 5 to 10 nm resolution with our best SPM probes, the initially measured average of 28 nm is well within resolution limits. As mentioned in the sample preparation procedure, the melting of fortuitous G-quartet structures through initial heating does not appear to generate a zero elapsed time, zero G-wire length. We speculate that growth of the multimers takes place extremely fast in a variety of growth conditions. Samples of the Tet 1.5 monomer taken from melting temperatures and imaged by the SPM have yielded no G-wire structures.

In Fig. 6b we find the common Poisson-like distribution of G-wire lengths from samples allowed to self assemble for several weeks. Under these conditions the mean length is approximately 73 nm in a potassium rich mixture. With each L_2 half-length of about 1.55 nm this corresponds to an average length multimer L_n of n = 47. Depending on growth conditions as much as 25% of the mixture remains as Tet1.5 oligonucleotide [4]. Each Tet1.5 oligo has molecular weight 3180 g/mole. The frequency distribution of G-wire lengths in Fig. 6b provides us with relative concentration of average G-wire lengths of 23%. If we assume this concentration reflects the relative amount of monomers tied up in the mean length multimer, then we can estimate the concentration of L_n for n = 47:

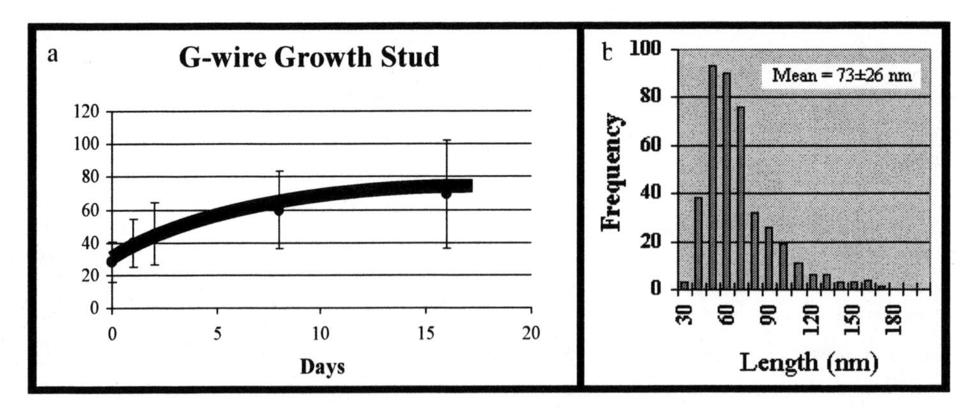

FIGURE 6. A measurement of the distribution of average G-wire lengths indicates a greater number of smaller wires (median equilibrium value of about 50 nm) over longer wires in Fig. 6a. N=400 from an image taken of G-wires grown for over a month. Error bars represent standard deviation. Note the presence of 30 nm in Fig. 6b wires even after growth for about a month. These short wires are present from essentially the very first measurement during the growth process, as seen in Fig. 6b.

$$[L_{47}] = (0.23)(0.75\text{mM})(3180\text{g/mole})/(286200\text{g/mole}) = 0.0019\text{mM} \quad (5)$$

Thus the equilibrium constant is:

$$K_{eq} = [L_{47}]/[G_4T_2G_4]^{48} \approx 10^{26} \quad (6)$$

The estimated equilibrium constant greatly favors the growth of the wires.

The fact that G-wires appear to self-assemble into smaller average lengths is challenging for the purposes of determining their macroscopic electronic characteristics. This is, large wires facilitate macroscopic electronic characterization. Curiously, diluting G-wire DNA samples with either growth or imaging buffers does not appear to greatly affect the average length of the molecules. From a thermodynamic growth standpoint, dilution should result in disassembly of the G-wires. However, dilution of samples to 1/1000 of their initial concentrations only reduces the density over the mica surface. The reason for this behavior is still being investigated.

G-wire Superstructures:

Occasionally, as seen in Figs. 7-9, G-wire superstructures are observed. They have appeared in three different forms – helices, ribbons and tubes (G-wire DNA) [6]. Fig. 7 includes two examples of flexible G-wire loops, similar in shape and size to plasmid DNA. If not for the fact these structures retain a minimum diameter of 2.2 nm when imaged in dry air, i.e. the diameter of G-wire DNA, contamination of the sample might be suspect. Double stranded DNA collapses on the surface of mica due to interactions between the substrate and the DNA [4]. The four-stranded G-wire DNA has always maintained its integrity on the surface of mica in dry air. The G-wire loops might be kinetically stable because, by self-assembling into a closed structure, disassembly is discouraged. There is no "end" on a closed loop for the ladder building blocks to vacate. The conditions under which such long structures develop, when the surrounding solution contains only smaller 100 nm length G-wire DNA, is still under investigation.

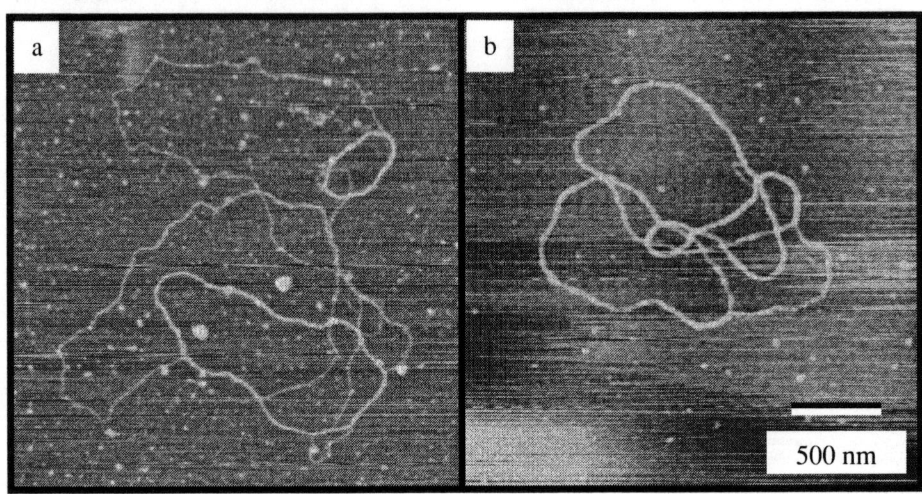

FIGURE 7. Interesting structures sometimes emerge including these examples of loops of single and multiple stranded G-wires. These results are found with numerous smaller G-wires surrounding the immediate vicinity as seen in the two examples provided here. We speculate that looped G-wire structures are stable against disassembly after dilution and rinsing by virtue of continuity. G-wires tend to maintain an average length much smaller than the contour lengths of the loops shown here. Multiple-stranding of G-wire DNA is common as seen in and Figs. 8 and 9, a result that is surprising in view of the strong electrostatic repulsion that is expected from the four strands of phosphate backbones on each G-wire. Vertical height range is 10 nm from black to white.

Flexible 1-D crystals have been found on occasion as show in Fig. 8. Fig. 8a provides a glimpse of a typical surface of mica inundated with G-wires approximately 100 nm long. Only the bright patch near the center of the image provides a hint of the longer structures found on the same surface a few hundred micrometers away. These spaghetti-like structures are several micrometers long, the kind of lengths of interest for electronic characterization. Two forms of evidence that these are G-wires are diameters over 2.2 nm or larger (larger presumably due to multiple wraps of the G-wires) and alignment of some segments of the structures with the mica substrate (arrowed regions on Fig. 8b). This alignment has been associated with the potassium vacancy-sites of the mica substrate (unpublished results). The sharp bends in these structures (arrowed regions), is suggestive of flexible one-dimensional crystals.

Evidence for more rigid structures is seen in Fig. 9a and 9c. Alignment with the potassium vacancies in the mica substrate is strongly evident in these pictures. The hexagonal arrangement of the potassium vacancies demands orientation at 60° intervals, as seen in these images. Unlike Fig. 8, the structures in Fig. 9 are rigid. Furthermore they appear to be multi-layered (Fig. 9b), as seen in the cross section image of Fig. 9a. The layers are integrals of half diameter of G-wires (about 1.1 nm). These structures might be examples of ladder G-DNA ($\mathbf{L}_1$) crystals as seen in Fig. 5.

Since these structures are much wider than the artificial broadening generated by finite tip geometry, with width could be explained by parallel packing of the ladder structures. A lattice match with the ladders and the potassium vacancy sites of the mica substrate (1.04 nm) would create a surface that the ladder DNA could also adsorb to, i.e. create multiple layers of ladder G-DNA. Note in Fig. 9c that flexible strands of G-wire DNA are also seen aligning in the same direction as the crystalline structures.

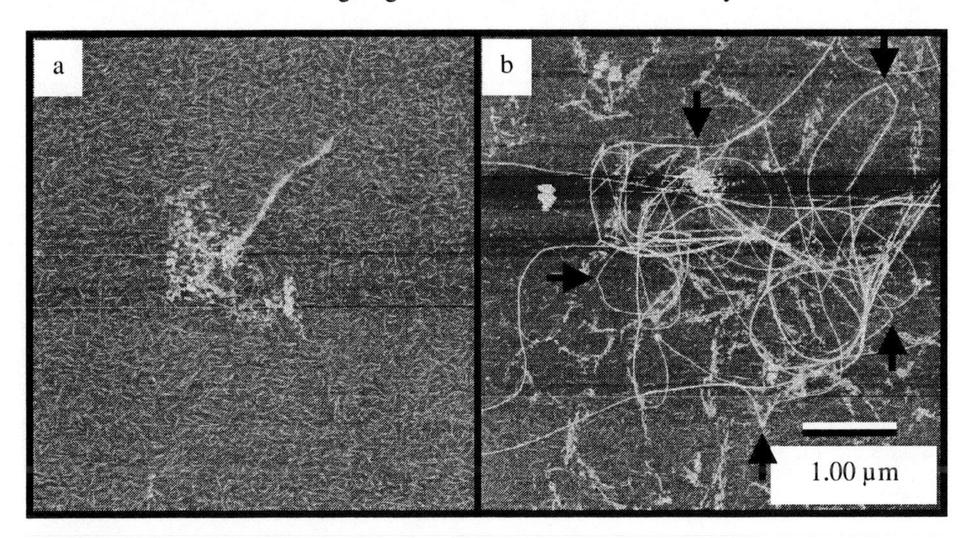

FIGURE 8. Two successive images from the same sample can be seen baring G-wires of the typical 200 nm length in the top image, and extremely long (hundreds of micrometers), spaghetti-like structures in the bottom image. Note the arrows in the bottom image indicating kinked regions of the spaghetti-like structures, and that the alignment of the straight lengths all are in the same direction. This orientation effect is a common feature of G-wire DNA (unpublished data). The diameter of these long structures (as measured by the height above the surface) is measured to be integrals of 2.2 nm. The latter is the diameter of the G-wire DNA as measured by SPM and other techniques [1,2]. This combination of evidence suggests that G-wire DNA under the right conditions can form one-dimensional crystals.

G-wire DNA makes an interesting candidate for molecular nano-circuitry because of the long lengths and narrow dimensions. Rinaldi et al. [7] undertook experimental investigation in which ribbon-like guanine structures (presumably like Fig. 8 below) were part of a metal – semiconductor – metal photodetector. I-V curves indicated “striking” semi-conductor behavior. Calzolari et al. [8] subsequently made first principle theoretical calculations that tube-like G-wire DNA should have semi-conductor-like behavior. Future research in our lab will focus on determining the conditions needed to create long wires and molecular networks.

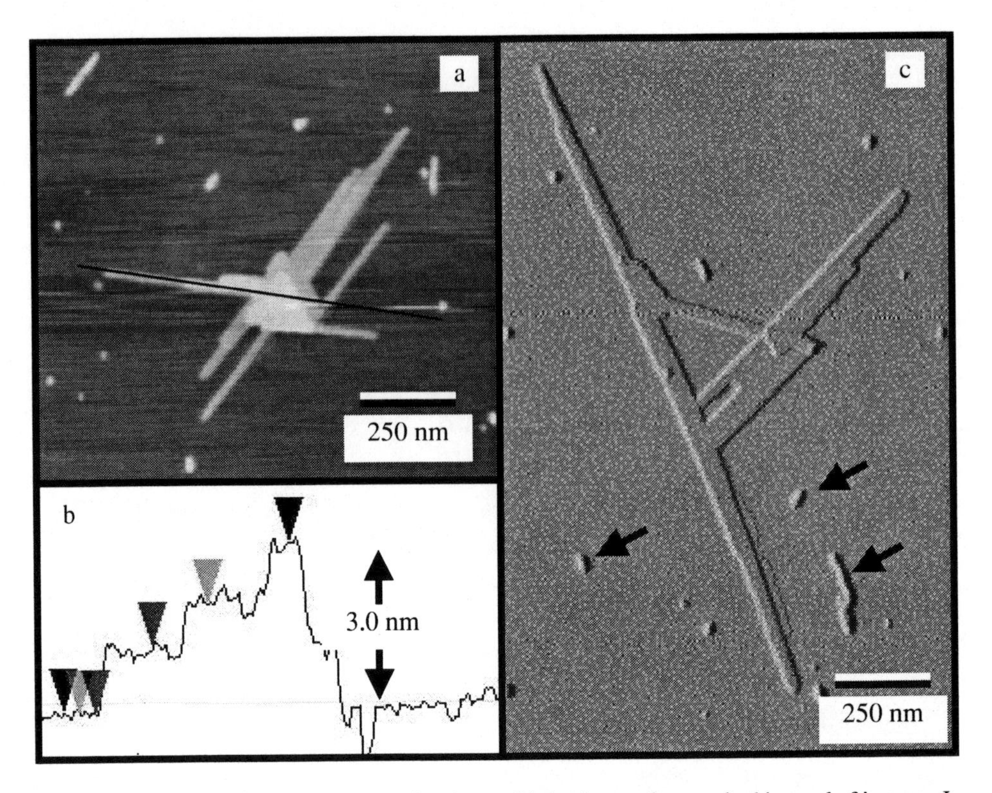

FIGURE 9. More evidence of one-dimensional crystal behavior can be seen in this panel of images. In Fig. 9a one-dimensional crystals are seen oriented at three 60° angles, orienting themselves with the underlying potassium vacancies in the surface of mica (unpublished data). Contact AFM image taken in dry air with a vertical height scale of 10 nm from dark to light. The black line through the crystal is a cross section described in Fig. 9b. This cross section indicates the different heights of the crystals, which form three distinct layers of heights 1.4 nm, 2.2 nm and 3.6 nm tall (each ±0.2nm). The height of the middle layer is exactly the average diameter of G-wire DNA as observed by SPM techniques, and the bottom layer is approximately 1/2 this diameter, the top layer is 3/2 this diameter. Fig. 9c is a deflection image example of the purported crystalline of G-wires, indicating the long lengths that these structures can attain (over a micrometer long). Note the arrowed flexible G-wire DNA segments in Fig. 9c and their alignment with the crystals, evidence of the lattice match with the underlying potassium vacancies of mica, and indirect evidence of the similar nature of the crystalline and flexible structures on the same surface.

ACKNOWLEDGMENTS

Undergraduate research assistant participating in this project: Tamieka Armstrong, Nicholas Demers, Kristin Eccleston, Matthew Fletcher, Mellissa Holden, Peter Hulsey, Marci Luhrs, Bethany Rioux. We acknowledge the financial support from the Research Corporation, University of New England, and the National Science Foundation MRI grant DMR-0116398.

REFERENCES

1. Williamson, J.R., Raghuraman, M.K., and Cech, T.R., *Cell.* **59**:871-880 (1989).
2. Williamson, J.R., *Proc. Natl. Acad. Sci. USA.* **90**:3124-3124 (1993).
3. TMarsh, T.C., Vesenka, J., and Henderson, E., *Nucleic Acids Res.* **23**:696-700 (1995).
4. Muir, T., Morales, E., Root, J., Kumar, I., Garcia, B., Vellandi, C., Marsh, T., Henderson, E., and Vesenka, J., *J. Vac. Sci. Technol. A.* **16**, 1172-1177 (1998).
5. M. Heim, R. Eschrich, A. Hillebrand, H.F. Knapp, & R. Guckenberger, *J. Vac. Sci. Technol. B* **14**, 1498 (1996).
6. Gottarelli, G., Spada, G.P., and Garbesi, A., in Comprehensive Supramolecular Chemistry, edited by Atwood, J.L., Davies, J.E.D., MacNicol, D.D., and Vögtle, F., Pergamon, New York, 1996, Vol. 9.
7. Rinaldi, R., Branca, E., Cingolani, R., Masiero, S., Spada, G.P., and Gottarelli, G., *App. Phys. Letters.* **78**, 3541-3543, (2001.)
8. Calzolari, A., Di Felice, R., Malinari, E., and Garbesi, A., *Appl. Phys. Letters* **80**, 3331-3334, (2002).

APPENDICES

Program

Thursday, May 23

19:30 Get-together
(Scala Restaurant- Intershop Tower, Downtown)

Friday, May 24

09:00 Opening
(IPHT at the Campus Beutenberg)

09:10 **N. Seeman** (New York University)
DNA Nanotechnology
09:35 **C.Dekker** (Delft University)
Single-Molecule Electronics, from Carbon Nanotubes to DNA
10:00 **M. Taniguchi** (Osaka University)
Development of electronic devices based on DNA

10:25 ---- coffee break ----

11:00 **W. Fritzsche** (IPHT Jena)
A construction scheme for a SET device based on self-assembly of DNA and nanoparticles
11:25 **C. Keating** (Penn State University)
DNA-directed assembly of metal nanowires
11:50 **K. Williams** (Delft University)
DNA-mediated Self-Assembly of Nanodevices

12:15 ---- lunch break ----

13:30 **M. Washizu** (Kyoto University)
Molecular Manipulation of DNA and Its Applications
13:55 **F. Bier** (Fraunhofer Inst. Potsdam)
Nanometer addressable lateral surface structuring by use of nucleic acids
14:20 **O. Harnack** (Sony Stuttgart)
Self-Assembly Driven Fabrication of Metallised DNA Nanowires and Their Electrical Properties

14:45 ---- coffee break ----

15:15 **M. Mertig** (University Dresden)
Biomimetic fabrication of metallic nanowires and networks
15:40 **S. Diez** (MPI Dresden)
Handling and Manipulation of single DNA-molecules by Kinesin driven microtubules
16:05 **C. Niemeyer** (University Bremen)
Semisynth. DNA-Protein Conjugates: Synthesis, Characterization and Applications in NanoBiotechnology

Program

(Friday, May 24)

16:30 Poster Session

K. El-Salam (Agricult. Res.Center, Egypt)
Effect of Gel Matrics on Characterization of F. oxysp. and F. oxysp. f.sp vasainvectum by RAPD Analysis

L. Franca (Munich University)
The Combination Of AFM And Laser-Based Microdissection As A Tool For Molecular Biology

L. Lie (Newcastle University)
On-chip solid phase DNA synthesis on semiconductor

R. Möller (IPHT Jena)
Surface-immobilized DNA constructs probed by electrical resistivity

S. Patole (Newcastle University)
DNA synthesis on silicon: An STM study

A. Sondermann (IPHT Jena)
Assembly of G-quartet based DNA superstructures (G-wires)

G.-J. Zhang (IPHT Jena)
Studies of an optical detection of DNA constructs based on nanoparticles and silver enhancement

G. Zuccheri (University Bologna)
A Direct Measure of the Ergodicity of the Small-Scale Dynamics of DNA

17:00 Orchid-Excursion

18:30 *Thuringian Bratwurst* - Dinner

Saturday, May 25

08:30 **B. Samori** (Bologna University)
Recognition of the DNA sequence by an inorganic crystal surface

08:55 **A. Pike** (Newcastle University)
Integrating DNA With Semiconductor Materials

09:20 **J. Burmeister** (Bayer AG)
SNP Analysis by Direct Electrical Detection

09:45 **P. Nielsen** (Copenhagen University)
Structure and Recognition Properties of Peptide Nucleic Acids

10:10 **G. von Kiedrowski** (University Bochum)
Nanobots - just science fiction?

10:35 ---- coffee break ----

11:00 **J. Vesenka** (College of New England)
Construction and Examination of "G-wire" DNA

11:25 **P. de Wolf** (Veeco Instruments) ***cancelled***
Scanning Probe Microscopes for Imaging and Manipulation of DNA

11:50 **M. Guthold** (Wake Forest University)
Novel Methodology to identify single aptamer molecules

12:15 **A. Woolley** (Brigham Young University)
DNA Alignment, Characterization and Nanofabrication on Surfaces

12:40 End of the scientific program

12:45 Lunch

Workshop Notes

by James Vesenka

Friday, May 24, 2002

9:10 Ned Seeman (New York University) DNA Nanotechnology: ned.seeman@nyu.edu Bottom up control of nano architecture. DNA as a molecule for making things – hard to make things that are linear in the topological sense. Branched junctions are created by minimize sequence symmetry. Sticky-ended cohesion: affinity, hydrogen bond sticky ends and then ligate to double bond. Combine branched junctions with sticky ends to make 2-D lattices. Objectives, architectural control, design molecules to assemble into ordered arrays, Nanomechamnical Devices, Self-Replicating systems. A new suggestion for producing macromolecular Crystals, make 3-D box, with macromolecules inside. Method for Organizing Nanoelectronic Components. Bruce Robinson, N. Seeman Protein Engineering, **1**, 295-300, DNA lots of useful features. Geometrical Constructions, Cube Junghuei Chen, Truncated Octahedran Zhang. Chen, Seeman 1991 Nature **350**, 631-633, Zhang, Seeman 1994, J. Am. Chem. Soc.? 1661?. DX Isomers. Fu, T.J. 1993 Biochemistry **32**, 11-3220, Lix YANG QI J. Am. Chem. Soc. **118**, 6121- (1996). Winfree, Nature **394**, (1998). Triple Crossover Molecules LaBean, J. Am. Chem. Soc **122**, 1848-1860 (2000). A 3-D TX Lattice. Special 1-D arrays like Six-Helix Bundle hoping for 2nm hole down the center. Mao, Nature **407**, 493-496 (2000), A Cumulative SOCR Calculation. DNA nanomechanical devices. Mao Nature **397**, 144-146 (1999). A device based on B <-> Z transition. Yan, Nature **415**, 62-65 (2002). Able to see PX-JX_2 Device with gels and SPM. Paranemic Cohesion Xiayang Zhang. References above probably available on Seeman's web site, http://seemanlab4.chem.nyu.edu

9:35 Cees Dekker (Delft University) Single-Molecule Electronics, from Carbon Nanotubes to DNA: http://www.mb.tn.tudelft.nl Carbon nanotube molecular wires. DNA for molecular electronics? Aviram, Mol. Cryst. Liq. Cryst. **234**, 13 (199?) Metalic (non-chiral molecules), semiconductor (chiral molecules). Need Low T STM to see atoms. Widldoer, Nature **391**, 59 (1998), Odom, Nature **321**, 62 (1998), Venema, Appl. Phys. Lett. **71**, 2629 (1997), Lemay et al., Nature **412**, 617 (2001), wave functions of electrons in nanotubes. Tans, Nature **386**, 474, (1997). , Tans, Nature **393**, 49, (1998) single-molecule transistor at room temperature based on a semiconducting nanotube. Chen JACS (2002). Yao, Nature **402**, 273 (1999). Posstma, Science (2001) (both single electron devices) Bachtold, Science (2001). DNA Electronics? Fink and Schonenberger (1999), Kasumov (2000) both highly conducting DNA. Braun Science (2001) DNA is an insulator. Bezryadin APL, (1997) looked at 10 nm polydGC. Storm, APL December (2001) ≈100 nm spacing between electrodes, <1pA at 10V, no conductivity, even with electrostatic gating. R>10 TW. Appears to be a fabulous insulator in a DRY state. Might be a conductor in a wet state. Speculates that short segments of DNA, because of their structural properties, connected to nanotobes. Zinc ion complexation might make DNA conductive (Brown University).

10:00 Masateru Taniguchi (Osaka University) Development of electronic devices based on DNA: Very good references on his abstract. Molecular devices with DNA. Network structures of polydG and polydC, gold coat over DNA and measure through gold-coated tip, find TeraW resistance. Sinh(aV) -> Polaron hopping. Using nanoscale FET experimental design. PolydA polydT DNA acts as an n-type semi-conductor. PolydGpolydC acts as p-type semiconductor. Doped DNA may work better. Electric field applied to DNA and with incident light might induce electron hopping. Nanopattering on Si substrate.

11:00 Wolfgang Fritzsche (IPHT Jena) A construction scheme of a SET device based on self-assembly of DNA and nanoparticles: Top-Down approach, easy interface but miniaturization is problematic. Bottom-up construction, can be very small, but interface is problematic. SET conventional, looks like an FET (gate, drain, required at lower temperature). Problem is structures are too big, possible solution, nanoparticles. Interface is problematic, can use SPM tips, statistical, nicest approach is electrostatic Bezyadin, APL **71**, 1273-1275 (1997). Can attach biowire through chemical binding between electrodes. Still need to position nanoparticle island. Attached either through thiol-modified Gold DNA, Amino modified DNA (silanized surface), non-specific charged gold surface. Flow alignment to control molecular orientation over the chip. 10-20% of the gaps had DNA between two surfaces. Positioning by in-situ hybridization (low efficiency), insert after restriction enzyme, or triple helix attachment (bind not as strong). Would like to nanoposition the particles (do not currently have this capacity).

11:25 Christine Keating (Penn State University) DNA-mediated Self-Assembly of Nanodevices (keating@chem.psu.edu)**:** 30-300 nm diameter wires. Trying to address challenges of assembly by two phase interface. Micropores from membranes are metallized on the back side with silver, alternate electrolyte, and then remove silver, and dissolve membrane. Can make all kinds of wires (e.g. cobalt which is magnetic), striped, etc. Striping can be read optically (e.g. reflectivity difference between gold and silver) at 430 nm sliver reflects, 600 nm both gold an silver reflect. Cool because this is a non-fluorescent method for particle encoding, applications Nicewarner-Pena, Science, **294**, 137-141 (2001). Just about as conductive as bulk gold ($2.2x10^{-6}$W-cm), though resistance decreases with diameter. Made a switching device with magnesium? Simplest nanodevice is a 2-D cross-point array. 10 parallel nanowires. Nonspecific binding accounts for 25% of wires. Mbindyo Adv. Mat. (2001). Problem is that assembly reaction is compounded by fast sedimentation. Assembly at the Liquid-Liquid Interface (2D rafts) by adding cyclohexane to the surface of the solution. DNA assembly at the interface. However, DNA hexane is not a good combination. Advantages, high concentration, 2D structures. Aqueous-aqueous interface (PEG polyethylene glycol and Dextran). Known to biologists. Stable up to 95°C. Problem, nanowires modified with thiolated DNA irreversibly aggregate at the interface. Non-complementary interaction stays apart, complementary aggregates.

11:50 Keith Williams (Delft University) DNA-mediated Self-Assembly of Nanodevices: Nanotubes attached to DNA. PNA-derivatized carbon devices. Motivation – high density, hands-free assembly, Leon Adelman "We steal from the cell... (Scientific American, 1998)." Tools of Biotechnology are used. Have made thiol-linked gold nanoparticles, but not a regular evidence of ordering. Want directional assembly, encode assembly instructions, error correction and/or re-configurability, replication, massively parallel assembly. Propose to use DNA to mediate and encode assembly of nanotubes. Nanotube FETs by Avouris and Bachtold Science, Nov 2001. Problem is that success depends on chance positioning, not all tubes are semiconducting, determined by the chiral indeces nm. Tubes with similar diameter can be wildly different conductors. Non-stick substrates, need to work in aqueous environment. Attach tube through PNA (see Peter Nielson's Talk for details). DNA provides unmatched molecular recognition, DNA is re-configurable. DNA must be attached to tubes, must be solubilized, need docking (connections) through lithographic limits. DNA attachement – carboxylate tubes at the end of the tube, then either covalently link adducts (thiols, amines) or hydrogen bond adducts such as DNA. PNA-DNA duplex (Nielsen and Haaima, 1997). PNA has uncharged backbone, free terminal NH_2 spacing of bases is almost identical to that in dsDNA, 20mers available. Soluble in water/DMF and many other solvents, T_m is higher per base/pair. Attached DNA-tube to mica through $MgCl_2$. Challenges to remain, efficiency of creation (need to purify). Can micellize nanotubes to obtain single walled nanotubes. Electrode attached DNA complementarily attaches to PNA-nanotubes. If insufficient conductivity then we add extra electrodes.

13:30 Masao Washizu (Kyoto University) Molecular Manipulation of DNA and its Applications: No spatial resolution in conventional solution based biochemistry. Nanotechnology based resolution involves micro-fabrication, micro-positioning, electrostatic manipulation, optical tweezers, etc. Electrostatic stretching and positioning of DNA. DNA polarizes under 1MV/m 1MHz field, DNA stretched by electrostatic force until one end touches and permanently anchored on aluminum electrode. Showed images with lambda phage DNA. Molecular Surgery of DNA with Enzyme-Immobilized Particle. Can cut with SPM tip, but no details of the cut. Instead use DNA-enzyme attached to nanoparticle that is move with laser tweezers. Works if DNA is free, but not if tied to the surface. Developed a floating-potential Electrode System that allowed the DNA to float freely. Can see it is free by fluid flow. Moved a particle over a stretched DNA to see it move the DNA about. Attached particle with non-specific DNAase and cut immediately. Next example is HindIII-bead, scanned along the DNA and then cut when the specific DNA sequence is found. Was able to characterize by examining lengths of cut DNA products. DNA-RNA polymerase interaction pushed against stretched and immobilized DNA. Can see the polymerase moving along the DNA.

13:35 Frank Bier (Fraunhofer Inst. Potsdam) Nanometer addressable lateral surface structuring by use of nucleic acids: Typical photolithographic surface are micron sized, want to work with DNA because it is significantly smaller. Use spacers (50nm) to keep biomolecule from the surface. M13mp17 Used 7223 bp (2 μm). Stretch in electric AC fields will orient the DNA (see Washizu's talk). Single molecule bridges, but no vectorial direction. So, label with PNA probes, can intercalate into the strands and can see different sized probes. Use oligo-tags and "micro-paint" but not very reproducible. Photochemical coupling with high special resolution, but not sharp? Containment in a conductive polymer, vectorial immobilization. Growth of electro-polymer on IDE (electrodes) after 3 potential cycles. Can see beads flopping around attached to the DNA. Use SPM to detect the DNA, but not useful because surface is too rough. Would like to apply NSOM or perhaps MFM. Tried bulk studies with different proteins such as telomerase. Also showed this could be used with PCR. Application: "Nanoarrays" Fluid volume of 1 fl (10^{-15} liter), content of single cell. Lots of interesting possibilities.

14:20 Oliver Harnack (Sony Stuttgart) Self-Assembly Driven Fabrication of Metallized DNA Nanowires and Their Electrical Properties: Bare DNA show high resistivity. Approaches for DNA Metallization, Ag, PD, Electrostatic binding of nanoparticles, Pt metallization. Gold nanoparticles on negatively charged DNA. Electro-less plating improves conductivity. Calf thymus DNA spin coated onto mica, attached with nanoparticles, and decorate by plating. Currently 30-40 nm in width. SEM pictures of wires attached to electrodes. Can get alignment perpendicular to electrode surface by fluid flow. Resistance is much smaller than DNA alone. Measured the resistance of a single metallized DNA wire. Cut wires and measured resistance change of parallel network to extract single wire, 2.3KW for the wire, bulk resistivity of $3.6x10^{-5}$Wcm. Stability against high bias voltage. As wire fails the resistivity goes up. Heat induced electromigration. Possible binding mechanisms, hydrogen bonds. Further steps: single-wire IV curves, understanding of templating mechanism, understanding of aging effect of NP solutions.

15:15 Ralf Seidel (University Dresden) Biomimetic fabrication of metallic nanowires and networks: Metallize, integrate into microscopic arrays, template fabrication using molecular motors. Metallization, bind $PtCl_4^{2-}$ to DNA bases (only a few percent of Pt bind), reduction (DMAB) of Pt complexes, growth of Pt clusters. Measure kinetics of DNA activation through absorbance at 600nm. Binding is faster as time proceeds. Seidel, Surface and Interface Analysis **33**, 151 (2002). AFM images show Pt clusters in time. High resolution TEM still show some gaps. What is microscopic metallization process? Stabilization of the first-formed Pt-Pt bonds by ligand-to-metal electron donation mechanism solved by molecular dynamics. Influence of the guanine-cytosine base pair content on the kinetics of cluster formation decreases with activation time. Growth of gold wires from DNA with Pt nucleation centers, measured nanoscale Pt metallization provided ohmic response. Interconnection of microscope Au electrodes, functionalized DNA attachment to gold electrodes, use hydrodyanmic flow to bridge the two electrodes. Also use biotin-streptavidin. Results of change in flow between two flips between the two electrodes. Also showed the beads easily trapped between two tied down beads, linked to DNA. Biomimetic materials synthesis borrows from biology, cellular mechanisms, and biomolecular templates.

15:40 S. Diez (MPI Dresden) Handling and Manipulation of single DNA-molecules by Kinesin driven microtubules: Richter et al., Appl. Physl. Lett., **78** 536 (2001). Motor proteins and cell motility like myosin attached to dyneins, actin filaments in muscle, microtubules sperm and cilia. Microtubules are non-covalently polar polymers, functions for cytoplasmic organization, cell division, organelles. Speed of microtubules is 0.8 μm/s (100 steps/s), energy source, ATP hydrolysis. Nice video of kinesin, attached to glass substrate, move microtubule shuttles within the clover slips region. Can guide microtubules on tracks. Objective is to build nanoeletrical networks with motor proteins and DNA – use DNA attached to microtubules (MT) that will be moved by kinesin motors to stretch the DNA out as it wraps around attached gold molecules. The DNA can then be used as a template for metallization, etc. DNA is biotinilated, MT is also biotinilated, streptavidin (SA) is used to connect DNA to MT. Optimize the environmental conditions – microtubules and motor proteins like high ionic strength, labeled DNA likes low ionic strength. DNA bound to MT requires care not to saturate SA with only MT-biotin. Used fluorescence dye to monitor both MT (red) and DNA (green) motion. Demonstrated, pick-up, transport and drop-off of DNA. Eventually build up nanostructures.

16:05 Christof Niemeyer (University Dortmond) Semisynth, DNA-Protein Conjugates: Synthesis, Characterization and Applications in NanoBiotechnology: cmn@chemmie.uni-dortmund.de. References in abstract. Once again top-down like photolithography or micro contact printing stops at about 1μm, bottom-up organic synthesis, self-assembly to a few nm. Biomolecules fill in the gap from 1-1000 nm. Concept supramolecular chemistry w/ biomolecules. Biotinilated DNA, attached to SA attached to more biotinilated DNA. Immuno-PCR, 1000 fold enhanced detection of single antibody antigen detection by duplicating the attached DNA. Use this technology to detect extremely low concentrations of rVISCUMIN anti-cancer drug, increased sensitivity by a factor of 4. Nanoparticle networks, streptavidin or 5nm Au Colloids and DNA supercoiling seen in AFM images, might be used for ionic switching of nanoparticle networks. Current SA-biotinDNA is randomly assembled. DNA streptavidin nanocircles. Soft Materials calibration standards. Hapten-modified nanocirlces. Immuno-PCR competes for analyte of binding molecules. DNA microarrays: Dendrimer-modified glass surface. ZB Centers, conventional surface chemistry, can be easily regenerated. Covalent DNA-STV conjugates. DNA-Directed Immobilizaiton (DDI). DNA-arrays -> protein arrays, enzyme microarrays, DDI-Kits for protein arrays. DDI of gold nanoparticles for detection of nucleic acids. DNA-Directed Nanocluster Assembly. Hunting for multi-enzyme constructs. Slides changed too fast for me to get details.

Poster Session from 17:00-17:30, many nice presentations-see abstracts for details, excursion thereafter to see orchids.

Saturday, 24 May 2002

8:30 Bruno Samori (Bologna University) Recognition of the DNA sequence by an inorganic crystal surface. First evidence of this phenomena head-to-head and tail-to-tail palendromic series. Were interested in sequencing structural and functional elements. Motion of linear DNA molecules on mica (assumed to be imaged by tapping mode in buffered media). Global shape of the molecule did not change very much. Locally the chain did equilibrate. Only local dynamics was ergodic. An equilibrium statistical ensemble of the shape of DNA molecule is needed. Because of the ergodic process it is okay to take an ensemble of images in air, which are much easier and quicker. Orientation of a molecule on a surface through the use of palendromic dimers, which is to say identical from the center of the DNA to ends so that topographic information could be taken from both ends. Information on the curvature was taken at both ends. Average from 1000 profiles provide local curvature versus fractional position along the chain. All the sequence-directed properties mapped along the chain should reflect the binary symmetry of the molecule. By reversing the dimer the local curvature map changed. Suggests that mica changes the curvature of the dimer? The most curved segment of DNA when prepared two palindromic dimers (also head-to-head and tail-to-tail) of the strong curvature also gave phases 180 opposite. Explained in terms of adenine rich face and thymine-rich face of the palindromic molecule. "S"-like shape provides more A-rich or T-rich face which is not true for "C" shape. Recognition of the DNA sequence by the mica surface? If mica interacts the same way with different DNA sequences a dimeric molecule will behave differently between S and C shapes. Curvature for S and S* were out of phase by 180 as expected. Face that is exposed is a T-rich one. Classification of the forms assumed by the molecules on the surface was provided, shows that that the T-rich population is preferred. Take this to infer that an inorganic mineral surface is able to recognize the sequence of DNA.
Question: Does the phosphate backbone interfere with the alignment process?
Answer: Evidence is strong for base alignment but understanding is incomplete, possibly due to minor groove subtleties which are related to T-rich region.
Question: Tried any other surface than mica?
Answer: No, just changing the teathering ion (Mg to ???).
Question: Time needed to achieving geometric equilibrium is about one hour on mica?

Answer: Statistics makes miracles.

8:55 Andrew Pike (University of Newcastle-upon-Tyne) Integrating DNA with Semiconductor Materials. Gordon Moore, density of silicon device is going to double every two years, only slowed down recently. Interested in etching of surfaces and the exposed silicon hydride surface and the step terraced surface. This surface can be functionalized by a variety of organic interactions (Chitwood). Examine with FTIR spectroscopy. Si-H undergoes alkylation and does not change very much under STM. Surface is modified to bring OH to surface. Pike, Angew. Chem. Int. Ed. **41**, 615 (2002). Boil chips to functionalize, attached to DNA. Nucleotides are cleaved and compared with purified DNA spread to show that the two are the same thus the cleavage is primarily at the base. DNA is hybridized onto UV patterned surface, seen by radioactive spreads. Quantitation of immobilized DNA. STM of DNA-modified Si (111) surfaces. See bundles of duplex DNA that tie up themselves and lie down on the surface. Hybrid biomolecules: modified DNA i.e. red-ox active nucleosides (A. Pike Chem. Eur. J. 2002 in press). Integrate this result into silicon to examine DNA based molecular electronics (no details yet). Deposition of polypyrole on DNA-silicon electrodes. Silicon-hydride, alkylation, DNA modified, polypyrole develops only in the DNA regions thus creating a conducting polymer surface.
Question: What are the real advantages of immobilizing DNA on silicon rather than thiol-gold?
Answer: We are using Silicon as electrodes. Using it because it is the electronics industry standard.
Question: Done any IV curves on the polypyrole surfaces?
Answer: No, haven't been able to get good STM images yet.
Question: Ferrocyne work an improvement over existing procedures?
Answer: Yes, not as harsh a treatment.
Question: Try to bind monomers onto surface and hybridize DNA onto afterwards?
Answer: No, have not looked at modifying bases themselves.

9:20 Jens Burmester (Bayer AG) SNP Analysis by Direct Electrical Detection. Market view of why we want to look at direct electrical detection. There is a huge growth worldwide because of the emergence of new technologies. Genetic tests, infectious disease, industrial and clinical trials. Single Nucleotide Polymerism (SNPs), 99.9% identical human genome, SNPs are most common types of genetic variations. Each individual has 4 million SNPs. Can be used with disease association and personalized care testing. Point-of-Care (POC) SNPs. Easy to use, fast results, minimal sample preparation, sensitivity, specificity, inexpensive handheld detection device, cost per test, flexibility. Insulating surface, hybridize surface (e.g. gold label), then attach gold to make quick detection by completing a circuit. Biofunctionalization, silicon, silination (Aptes), functionalize molecules to attach to surface and target molecule. Functionalized molecules were biotin and enhanced SA-gold, gold array showed 100% conductivity for matched groups and no activity for mismatched targets. SNP discrimination on a chip uses DNA complementary sequence. Tested with single stranded probes and targets with only single change in the sequence. GC has high pacing of colloids, AT next, GT then AC is least. GC and AT had 100% electrode conductivity, GT and AC had zero. Fluorescent readout would only give degrees of intensity. SNP in unlabeled target: Electrical detection shows normalized conductivity as a function of enhancement time that shows could discrimination in about 11-13 minutes. Imagine straight forward applications such as PCR products. Perspective: DNA detection and gold, a whole slew of possibilities (microscopy etc.), SNPs detection adds another tool to the list.

Question: Reason for threshold is an autocatalytic effect. How to further increase the sensitivity? Conceivable to show two oligos?
Answer: Could have used branch DNA to introduce numerous gold particles. For now just testing, but believes there are many ways to boost efficiency of these systems.
Question: Nanopoint detection, can it become on-line detection method?
Answer: Interested in pursuing this path but not undertaken because it is time consuming.
Question: Need stringent washing protocols.
Answer: Short hot washes.
Question: How does it work when you go to the doctor? Need to get whole genome to get relevant piece?
Answer: That's a problem, if you use PCR it amplifies up the tiny sequence you can interrogate. Not yet solve but is well known to the research community.
Comment: Not all SNPs are the cause of disease, quite a few are silent. Of the million or so identified some are probably the result of poor sequencing.
Question: When do you start to lose sensitivity?
Answer: We grow the colloids up to 200 nm, would assume that we can get a ten-fold diameter increase.

9:45 Peter Nielsen (University of Copenhagen) Structure and Recognition Properties of Peptide Nucleic Acids (PNA). DNA mimic, built on peptide chemistry, more closely related to protein chemistry. Can hybridize to DNA. But it is uncharged. Cannot make PNA as long as DNA, hydrophilic, reasonably soluble in water. Chemically and biologically stable. Chemical versatility and flexible, e.g. attach fluorophores, even combined PNA and DNA chemistry. Binding does not depend on ionic strength. Binds more strongly to PNA than other nucleic acids. Purine-PNA is much more stable the pyrimidine-PNA, but homo-pyridine-PNA triplexes are very stable. PNA adapts to the partner, PNA-RNA is A form, PNA-DNA is B form, starts to deviate as more PNA is integrated into the structure, which alone is a long helix. Four binding modes: Triplex: homopyrimidinhe PNA (C-rich), Triplex invasion: homopyrimidine PNA, duplex invasion: homopurine PNA, double duplex invasions: Pseudo complementary PNA. Triplex invasion, Watson-Crick plus Hoogsteen pairs. Can only recognize A or G? Double duplex invasion is both Watson-Crick pairing, but since complementary the PNA can quench each other. This binding can take place and be stable if you have about 50% ATs. PNA properties: dsDNA binding: Helix invasion, binding/specificity, slow kinetics. Backbone-modified PNAs. Surface Immobilized PNA: glass/gold/thiol/strepatvidin/probe DNA, target DNA, fluorophore. Surface plasmon resonance spectroscopy is used to examine surface hybridization kinetics. PNA cannot be used as DNA because of its special properties. Transcription induced supercoiling. DNA is fixed in nuclear matrix, RNAP will rotate. As RNAP generates mRNA, the DNA is forced to rotate and supercoil. PNA anchored DNA Immobilization. DNA is a closed loop. Sequence specificity of targeting. Gel sequence indicates binding with both single, double mismatches as well as the target. Dexametasone assisted gene delivery. Problem is getting DNA to transvect into the nucleus. PNA can help by tagging the vectors with an attached reagent that helps to transport the gene into the nucleus. Was very successful.

10:20 Günter von Kiedrowski (Ruhr University Bochum) Nanobots – Just Science Fiction? Drexler (1987), nanorobots, self-replicating nanomechanical devices, replace organic chemistry by a kind of AFM-type manipulation of atoms to fabricate new structures. Utility "foglets" that self assemble into utility "fogs". What could be a nanobot? A replicatable array of modular functions in a defined 3D arrangement organized into stiff structures. Two design

principles for noncovalent nanoobjects. Tensegrity = integrity upon tension. For example, tetrahedron is stiff but cube can fold. Also maximal instruction. Seeman's 3D DNA nanostructures. Seemen's Strategy, to go from noncovalent to covalent structures. Our strategy, self-assembly into non-covalent structures. See abstract for references. Fast cooling yields nanostructures from Trisoligonucleotidyls. Slow cooling yield polymeric networks. Rapidly cooling as a means to apply kinetic control during noncovalent synthesis may lead to maximal instruction. How to synthesize a trisoligo from a bislinker creates both 3 and 3+1 arm building blocks. Use gels to identify different structures. UV melting curves indicate tetrahedron shows cooperative melting, so two-state model is not applicable because there is melting diversity in the individual duplex bonds. CD studies indicate B-DNA form. Self assembly of Trisoligonucleotidyls. Chemical copying of connectivity information yields 5'-trisoligos from 3'-trisoligo templates. The SPREAD procedure. Potential as a nanostructure preparation. Immobilize, hybridize, ligate, electrotransfer, then by flip-flop you amplify exponentially. Chemical translation of connectivity information. Gold cluster in nanobiotechnologies: TEM labeling, universal fluorescence quenchers for proximity probing, nanoscale antenna for single molecule heating. Nonotechnology needs heating steps. RUBigold principle and synthesis. What could be a nanorobot? Self assemble a naked scaffold in which functionalized junctions are included.
Question: Movie showed nanobot destroying bad objects in the cell.
Answer: An artist's example of a nanotbot killing a virus like a staph bug.

11:00 James Vesenka (University of New England) Construction and Examination of "G-wire" DNA: Interested in trying to characterize the growth process of novel, four stranded DNA, a.k.a. "G-wires". Examination of growth kinetics through free energy of dimerization and kinetic equilibrium. Normalized mean length of G-wires as a function of time provides long time constant, experiment is being re-examined. More interesting results are apparent one-dimensional crystals and orientation of G-wires on mica. The height of dsDNA on mica has been clearly explained in terms of the teathering ions. But the apparent one-dimensional crystals of G-wires behave strangely. Orientation on mica of shorter segments of G-wires appears to be a function of alignment with the next nearest neighbor repeat sequence on mica with the phosphate backbone of the G-wires.
Question: Slow versus fast cooling (Gunter von Kiedrowski) may affect self-assembly.
Answer: After seeing von Kiedrowski's talk I have to reexamine my growth protocols.
Question: Diffusion based control (square root of time dependence) kinetics?
Answer: Have obtained diffusion results, but time constant is ridiculously small. Problem is in growth protocol.
Question: Pt or Pd salts in growth medium?
Answer: Good suggestion, have only used magnesium and zinc till now.

11:25 Peter DeWolf (Veeco Instruments GmbH) Scanning Probe Microscopes for Imaging and Manipulation of DNA: No show, would have been interesting.

11:50 Martin Guthold (Wake Forest University) Novel Methodology to Identify Single Aptamer Molecules. Aptamers oligonucleotides which have a demonstrated capability to specifically bind molecular targets with high affinity ($K_D = 10^{-6}$ to 10^{-9}). E.g. Thrombin binding aptamer Bock, Nature **355**, 564 (1992), 15 nucleotide (nt), K_D = 25nM, structure, two stacked G-tetramers with each of the two strands having a dinucleotide loop. 200 known aptamers, 15-100 nt long. Ellington lab http://aptamer.icmb.utexas.edu data base find versatile target, peptides, proteins, organic molecules, inorganic molecules, may be useful for drug

discovery. Aptamers are found by SELEX: Systematic Evolution of Ligands by EXponential enrichment. Start out with a randomized library of 10^{15} oligos. Bind to target molecule attached to a solid support (like a bead). Select bound molecules, wash and elute, reduce number of oligos by 90%, but still have 10^{14} oligos to choose from. Undertake PCR, bind to target molecule under more stringent conditions, repeat process, 90% enrichment, repeat 6-12 cycles (about one year), if you are lucky, you get an aptamer by the end. Advantages, it works. Disadvantages, time consuming, tedious, difficult to check oligo pool in the early rounds, heterogeneous pool of aptamers (difficult to sequence). Proposed new method, a single (or few) cycle(s). Library of aptamers labeled with an acceptor fluorophore, target molecule which is labeled with a donor fluorophore, bind and wash, sp-FRET, donors unpaired emit green, unbound acceptors do not emit red light. But those that are close together provide fluorescence resonance energy transfer can be observed optically, specific binding is examined by AFM and FRET, align images, pick up with AFM tip, isolate, amplify and characterize. Ultimate Target – tumor cells. Her-2, a membrane protein, signals cell division. Currently at the proof-of-concept stage. Well known thrombine aptamer for initial studies. Examined specific vs. non-specific binding. Attached 190-nm bead through streptavidin-thrombin aptamer, bind and wash, blocked aptamer with 1000x thrombin, no non-specific binding. Know binding constant of thrombin aptamer. Half will be bound, half will not, will be proportional to concentration. Undertook dose dependent study binding and non-specific binding. Non-specific binding is still a problem. Next step, picking up a molecule. Image bead without disturbing the bead (tapping in air) then you need to pick up the bead, go to oxidizing environment. We use a nanoManipulator, based on Topometrix Explorer. Image is noisy so now we have picked up the bead, supposedly with the aptamer. Test out with single molecule PCR. DNA on AFM tip has been amplified by Thalhammer et al chromosome, Xu et al. plasmid.
Question: Are you going to image all 10^{15} possible aptamers?
Answer: No, problem is time, need to let all aptamers a chance to bind to the targets. Can image an area of 10000 μm^2, can see lots of beads. Perhaps use SELEX for some pre-selection.

12:15 Adam Woolley (Bringham Young University) DNA Alignment, Characterization and Nanofabrication on Surfaces. atw@byu.edu, chemwww.byu
Studying DNA on surface, easily characterized by AFM, carries genetic information, valuable to study surface interactions, high aspect ratio. Next generation probe tips. Silicon tips has R_c>10 nm. But nanotube (NT) has smaller radius, high aspect ratio, elastic buckling minimizes sample damage, terminally functional, reuse by removing end of contaminated NT, long life time. Fabrication, mechanical mounting – labor intensive. CVD growth directly on tips. Length optimization is nontrivial. Pick-up tips: scan across CVD wafer, only one tip at a time from CVD tubes. Holds by VDW interaction. Generate many tips from the same substrate. Adjust nanotube tip length by controlled surface Hafner J.Phys. Chem B **105**, 743 (2001). Advantages, lack of complementary strand, direct hybridization, smaller diameter than dsDNA. Disadvantages, lack of complement, intramolecule base pairing, previous effort not reproducible. Yokota et al. Anal. Biochem., **264**, 158 (1998), Aligning well extended ssDNA, poly-L-lys on substrate, add DNA, translate DNA droplet across surface, rinse, dry image. Three key experimental parameters. 1-10 ppm poly-L-lysine surface, so that DNA can bind positively charged surface. Low ionic strength DNA solution 10 mMTris, 1mM EDTA or lower, controlled droplet translation on the surface (Woolley, Nano Lett., **1** 345 (2001). Mechanism of DNA Extension. Surface Tension or Fluid Flow Frictional Drag, mostly likely the latter estimated at 7pN. 75% traceable contour, well extended lambda DNA on mica.

Higher concentrations of lysine (10ppm) incomplete stretching. Extensive study of sample preparation – lower concentration of lysine reduces concentration of DNA. Where we image with respect to trail determines concentration of DNA. Multi-step surface preparations, generate more sophisticated DNA arrangements, controlled deposition, test stability, orthogonally aligned ss and ds DNA, height of ds-DNA is 2x ss-DNA. Alignment depends on surface treatment 10 ppm perpendicular to drop motion, 1 ppm parallel to drop motion. Transition takes place around 4 ppm. Nanofabrication, surface DNA is a template for nanowire fabrication, electrostatically attracts metal cations, assembled cations can be reduced to metal atoms, multiple deposition/reduction steps build up nanostructures. AFM result for nanowire growth, treat with 0.1M$AgNO_3$ for 5 minutes, expose to light 3 minutes, rinse. Untreated DNA goes from 1.14 nm in height to 2.45 nm height after 3x $AgNO_3$ deposition, appears to be 1.2 nm height increase with Ag. Nanoscale positioning is controlled by macro-scale process.

Question: Do you protect the sample from slide?

Answer: Slide never touches the sample, about 1mm away.

Question: Resolution from nanotips?

Answer: 5-10 nm resolution, not a lot better than regular tips. Contaminants ruin Si tips, but NT tips extend tip lifetimes. Single wall nanotubes improve images, mostly multiwall.

Photos

Talks (*Friday, May 24*)

N. Seeman (New York)

C. Dekker (Delft)

M. Taniguchi (Osaka)

W. Fritzsche (Jena)

C. Keating (Penn State)

K. Williams (Delft)

Photos

M. Washizu (Kyoto)

F. Bier (Potsdam)

O. Harnack (Stuttgart)

R. Seidel (Dresden)

S. Diez (Dresden)

C. Niemeyer (Bremen)

Orchid-Excursion

Photos

Talks (*Saturday, May 25)*

B. Samori (Bologna)

A. Pike (Newcastle)

J. Burmeister (Bayer)

P. Nielsen (Copenhagen)

G. v. Kiedrowski (Bochum)

J. Vesenka (Univ.N.Engl)

M. Guthold (Forest Univ.)

A. Woolley (Young Univ.)

LIST OF PARTICIPANTS

F. Bier
Fraunhofer IBMT
bier@ibmt.fhg.de

J. Burmeister
Bayer AG
jens.burmeister.jb@bayer-ag.de

L. Capes
Motorola CRM
capes@crm.mot.com

B. Connolly
University of Newcastle
b.a.connolly@ncl.ac.uk

A. Csaki
IPHT Jena
csaki@ipht-jena.de

P. Cumpson
National Physical Laboratory
pjc2@npl.co.uk

C. Dekker
Delft University of Technology
dekker@mb.tn.tudelft.nl

P. De Wolf
Veeco Instruments
dewolf@veeco.fr

S. Diez
MPI of Molecular Cell Biology and
Genetics Dresden
diez@mpi-cbg.de

L. Eckardt
Ruhr-University Bochum
lars@orch.ruhr-uni-bochum.de

F. Eckstein
MPI of Exp. Medicine Göttingen
eckstein@em.mpg.de

K. El-Salam
Institut für Phytopathologie Uni Kiel
kelsalam@phytomed.uni-kiel.de

U. Feldkamp
University Dortmund
feldkamp@LS11.cs.uni-dortmund.de

Elisenda Ferrer
FRIZ Biochem
elisenda.ferrer@frizbiochem.de

G. Festag
IPHT Jena
festag@ipht-jena.de

A. Filoramo
Motorola CRM
filoramo@crm.mot.com

W. E. Ford
Sony International
ford@sony.de

L. Franca
Institut für Kristallographie und angewandte
Mineralogie München
lila@lrz.uni-muenchen.de

W. Fritzsche
IPHT Jena
fritzsche@ipht-jena.de

R. Gehrmann
University Düsseldorf
RuGehrmann@aol.com

M. Guthold
Wake Forest University
gutholdm@wfu.edu

U. Haker
FRIZ Biochem
ute.haker@frizbiochem.de

O. Harnack
Sony International
harnack@sony.de

R. Hölzel
Fraunhofer IBMT
ralph.hoelzel@ibmt.fhg.de

B. R. Horrocks
University of Newcastle
b.r.horrocks@ncl.ac.uk

Andrew Houlton
University of Newcastle
Andrew.Houlton@ncl.ac.uk

T. Kaiser
Clondiag chip technologies
Thomas@clondiag.com

C. Keating
Penn State University
keating@chem.psu.edu

G. von Kiedrowski
Ruhr-University Bochum
kiedro@ernie.orch.ruhr-uni-bochum.de

J. Kobow
BioRegio Jena e.V.
kobow@bioinstrumente-jena.de

K. Gloddek
RWTH Aachen
kirsten.gloddeck@ac.rwth-aachen.de

L. Lie
Newcastle University
L.H.Lie@ncl.ac.uk

A. Lührs
DEWB Jena
alexander.luehrs@dewb-vc.com

S. Mönnighoff
Ruhr-University Bochum
sven@orch.ruhr-uni-bochum.de

D. Martin
University of Liverpool
davidm@liverpool.ac.uk

G. Maubach
IPHT Jena
maubach@ipht-jena.de

N. Matsuzawa
Sony International
matsuzawa@sony.de

M. Mertig
Technical University Dresden
mertig@tmfs.mpgfk.tu-dresden.de

R. Möller
IPHT Jena
moeller@ipht-jena.de

K. Naumann
Ruhr-University Bochum
kai.naumann@ernie.orch.ruhr-uni-bochum.de

Dr. C. M. Niemeyer
University Bremen
cmn@biotec.uni-bremen.de

P.E.Nielsen
University of Copenhagen
pen@imbg.ku.dk

M. Noyong
RWTH Aachen
Michael.noyong@ac.rwth-aachen.de

Filipp Oesterhelt
MPI BCB
filipp.oesterhelt@nanotype.de

S.N.Patole
Newcastle University
S.N.Patole@ncl.ac.uk

A. Pike
Newcastle University
a.r.pike@ncl.ac.uk

J. Reichert
IPHT Jena
reichert@ipht-jena.de

C. Reuther
MPI of Mol. Cell Biol. and Gen. Dresden
reuther@mpi-cbg.de

B. Samori
University of Bologna
samori@alma.unibo.it

A. Schulze
DEWB Jena
andreas.schulze@dewb-vc.com

N. Seeman
New York University
ncs1@SCIRES.ACF.nyu.edu

R. Seidel
Technical University Dresden
seidel@tmfs.mpgfk.tu-dresden.de

A. Sondermann
IPHT Jena
anett.sondermann@ipht-jena.de

K. Steinmetzer
Quantifoil Micro Tools GmbH
katrin@quantifoil.com

M. Taniguchi
Osaka University
tanigu32@sanken.osaka-u.ac.jp

T. Tomonaga
Mitsui & Co. Deutschland GmbH
Jhodgson@dus.xm.mitsui.co.jp

E. Tornquist
Karolinska Univ. Stockholm
elisabeth.tornquist@biosci.ki.se

Jens Tuchscheerer
CLONDIAG chip technologies
jens@cjondiag.com

J. Vesenka
Univ. of New England, Biddeford (MN)
jvesenka@une.edu

M. Washizu
Tokyo University
washizu@washizu.t.u-tokyo.ac.jp

M. Wevers
VDI
wevers@vdi.de

K. Williams
Delft University of Technology
K.A.Williams@tnw.tudelft.nl

A. Woolley
Brigham Young University, Provo (Utah)
adam_woolley@byu.edu

G. Zuccheri
University of Bologna
giampa@alma.unibo.it

G-J. Zhang
IPHT Jena
guojun.zhang@ipht-jena.de

AUTHOR INDEX